Special Forces in the
New Battle Against Terrorism

SHADOW
WARS

by David Pugliese

ABOUT THE AUTHOR

David Pugliese, a journalist with the *Ottawa Citizen* newspaper, has been writing about military affairs and the Canadian Armed Forces since 1982. His reporting on defence issues earned him a National Newspaper Award in 2002. Since 1990, he has also been a correspondent for the U.S. publication *Defense News*. Pugliese's freelance articles on military issues have appeared in more than a dozen publications, including *Armed Forces Journal International, Bulletin of the Atomic Scientists, Training and Simulation Journal, International Combat Arms,* and *National Defense* magazine.

NATIONAL LIBRARY OF CANADA CATALOGUING IN PUBLICATION DATA

Pugliese, David, 1957 -

Shadow wars: special forces in the new battle against terrorism

Includes bibliographical references and index.

ISBN 1-895896-24-X

1. Special forces (Military science) 2. Terrorism--Prevention

U262.P84 2003   356'.16   C2003-905445-4

Printed and bound in Canada

Esprit de Corps Books

1066 Somerset Street West, Suite 204

Ottawa, Ontario, K1Y 4T3

1-800-361-2791

www.espritdecorps.on.ca / espritdecorp@idirect.com

From outside Canada

Tel: (613) 725-5060 / Fax: (613) 725-1019

AUTHOR'S NOTE

They have been called commandos, super soldiers and shadow warriors. They are members of special operations forces – or SOF – the most elite, most skilled and certainly the most enigmatic of military fighters. Since September 11, 2001, SOF troops from numerous allied nations have found themselves on the front lines of the war on terrorism.

Shadow Wars: Special Forces in the New Battle against Terrorism takes a look at the first two years of this conflict, and the dangerous and often secretive role played by SOF. While not attempting to encompass the entire fight against terrorism, it is hoped this book will give readers insight into some of the valuable contributions being made by special operations forces and into the evolution of their role as the rules and the very nature of global warfare change.

A number of organizations have provided photographs for this publication and should be thanked. Those agencies include: U.S. Special Operations Command (SOCOM), the United States Navy (USN); U.S. Department of Defense (DoD); U.S. Army; United States Air Force (USAF); U.S. Central Command (CENTCOM); Canadian Forces Combat Camera (Combat Camera); and the British Ministry of Defense (MoD).

I would also like to make particular mention of Emma Diffen of the Australian Defence Force (ADF) who graciously – at 4 a.m. Australian time – provided the necessary information for the use of the exceptional ADF pictures that are reproduced here.

In Canada, Julie Hallée at the Department of National Defence worked diligently to provide me with the JTF2-Afghanistan photographs for publication. Also playing a key role in that task was the team from the Office of the Information Commissioner of Canada, John Reid. Those individuals include: Dan Dupuis, Roy Hillier and Dave St-Pierre. Major James Simiana worked with the unit for the clearance of the other JTF2 photographs. Lieutenant General Michel Maisonneuve and Dan Dupuis cleared the way for the release to me of the JTF2 recruiting poster.

At the *Ottawa Citizen*, Editor Scott Anderson, Managing Editor Lynn McAuley and my assignment editor Bruce Garvey have always been supportive of my defence writing endeavors. I thank them for that strong support. My colleagues, Sharon Hobson and Jim Bronskill, have always provided sound advice on military and intelligence matters. Others who I wish to thank are Peter Holtby and Rob Webb.

Thanks, finally, are due to Scott Taylor, Katherine Taylor, Cathy Hingley and Julie Simoneau of *Esprit de Corps* who designed and laid out the contents of this book.

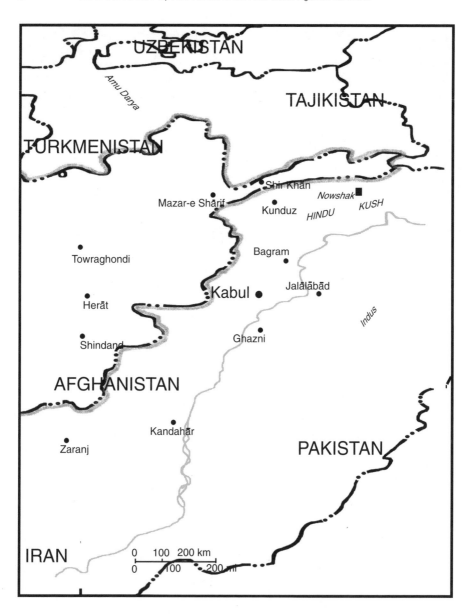

ABOVE: *Map of Afghanistan.*

OPPOSITE PAGE: *A U.S. Army Chinook helicopter carries SOF personnel during a mission in Afghanistan in 2001. (COURTESY ADF)*

A New Global War

1 "I'm going to fucking shoot you!" The American special forces soldier
barked out the threat and glared at TV journalists who were videotaping
the arrival of his fellow commandos at the Qala-i-Jangi fortress in northern Afghanistan.

The bearded operator, clad in a tan cargo vest and khaki pants and carrying an
AK-47, was clearly infuriated that the news media were documenting what was
supposed to be a covert mission. He demanded that the cameras be turned off
immediately and let reporters know the consequences if his orders weren't followed.

"I'm going to shoot you," he yelled again at German TV journalist Arnim Stauth.
"Put that in your notes!"

The U.S. commandos, most with beards and wearing sunglasses and baseball
caps, had stopped at the side of a road near the massive sand-colored fortress. The
Americans were climbing out of their mini-vans just as two white Land Rovers,
each with a couple of armed British commandos on the roof, pulled up behind
them. As soon as those soldiers spotted the journalists, they covered their faces
with scarves or their arms.

As the day wore on, an assortment of American and British special operations
forces, or SOF, gathered outside the fort. From the U.S. there were Delta Force,
Green Berets and Air Force combat controllers. The British commandos were from

the Special Boat Service, and there was speculation that members of that country's Special Air Service, who were reportedly housed at a nearby school, had also been called to the scene.

Inside the Qala-i-Jangi fortress, some 400 to 450 Taliban fighters were doggedly holding onto the sprawling complex, despite several days of pounding by U.S. fighter aircraft. The uprising had already claimed the life of a 32-year-old CIA agent, Johnny "Mike" Spann, who had been interrogating some of the captured foreign Taliban inside the fortress when they rose up and seized control of Qala-i-Jangi.

The former Marine Corps captain, who had been working for the spy agency's Special Activities Division, was to become the first American casualty in what U.S. President George W. Bush called "the first war of the 21st Century."

It was late November 2001, and the fighting in Afghanistan had been underway for six weeks.

Back in New York and Washington, cleanup crews were still sorting through the ruins of the World Trade Center and the heavily damaged Pentagon, the main targets of the September 11 attacks that had temporarily paralyzed the most powerful nation on earth and had later unleashed America's enormous military might on Afghanistan.

Just hours after hijacked aircraft had plowed into the Trade Center's Twin Towers and the Pentagon, the CIA had identified Saudi Arabian millionaire Osama bin Laden as the most likely culprit behind the attack. Bin Laden was the guiding force behind al-Qaeda or the Base, a Muslim terrorist network that had vowed to punish the U.S. for its long-standing support of Israel and its ongoing military presence in the Middle East. Al-Qaeda supporters had been behind the 1993 bombing of the World Trade Center and attacks on American embassies in Africa five years later.

Terrorists had been a global scourge in the late 1960s and early 1970s, hijacking aircraft, detonating bombs and taking over embassies. But the September 11 attacks were different. America had never been struck on such a large scale and with such ferocity by foreign terrorists on its own soil.

Bin Laden was operating from Afghanistan with the full support of the Taliban regime, which ran the country under strict Islamic law. His training camps were churning out hundreds of fanatics who, like those behind the controls of the passenger jets used in the September 11 attacks, wouldn't hesitate to die for their cause.

President Bush had vowed to bring in bin Laden dead or alive and to destroy his al-Qaeda network. The Taliban government would also be eliminated after ignoring American demands to turn over the Saudi.

What became known as the new global war on terrorism began on October 7, 2001, when the U.S. unleashed waves of cruise missiles and bombers to strike at Kandahar, Kabul and other Afghan cities. Two weeks later, U.S. and British commandos were operating inside Afghanistan with anti-Taliban forces, known as the Northern Alliance. By the last week of November, when the revolt at the Qala-i-Jangi fortress began, the Taliban regime was teetering on the brink of collapse and al-Qaeda was on the run.

Moving Taliban prisoners to the fortress, in fact, was supposed to be a gesture of reconciliation on behalf of one of the Northern Alliance's most successful generals. After the rout of Taliban forces near Kunduz, General Abdul Rashid Dostum had agreed to a cease-fire arrangement that would allow Afghan Taliban fighters to return to their families. Foreign Taliban, including those from the Middle East, Pakistan and Chechnya, would be taken to Qala-i-Jangi and eventually turned over to the United Nations.

The 19th-century fortress was an ideal compound in which to hold the prisoners. A sprawling complex just outside Mazar-e-Sharif, it served as General Dostum's headquarters and had several spacious parade grounds and courtyards dotted with trees, buildings and bunkers. An ornate white stone gate marked the entrance and the sand-colored brick walls of the fort were, in some parts, almost 10 metres thick. Its angled ramparts made the fortress the dominant feature on the surrounding plains.

As Taliban prisoners entered Qala-i-Jangi aboard transport trucks, Dostum's men searched the fighters for hidden weapons. A handful of grenades and pistols were found among those in the first vehicles. But with darkness falling, Northern Alliance guards waved through the last two truckloads of prisoners, failing to conduct any search of those men.

The next morning, on Saturday, November 24, that failure would have deadly consequences. One of the foreign Taliban produced a grenade he had hidden in his clothing and pulled the pin, killing himself and one of Dostum's senior officers. A few hours later, another prisoner killed himself and a Northern Alliance commander in the same way.

In response, Dostum's troops rounded up the men and locked them in several buildings throughout the compound to await interrogation by CIA agents the next day.

Those operatives, Johnny Spann and his partner, Dave Tyson, arrived at the fortress early Sunday to begin trying to determine which of the prisoners might be valuable sources of information on al-Qaeda. It is not known whether the two agents were told of the suicide grenade incidents the day before and the possibility that some Taliban prisoners might still possess hidden weapons.

Strangely, the two Americans decided to bring out a large number of prisoners for interrogation instead of questioning each individually. One by one, Taliban were bound by rope and seated on the ground in one of the courtyards as the CIA agents walked among them, singling out specific individuals for questioning.

Later, video footage of some of the interrogations was released showing Spann and Tyson in the midst of at least 40 enemy soldiers.

"I think you're a terrorist," Tyson said to one of the foreign Taliban.

The man, a Pakistani, protested and the CIA agent turned to Spann, who had an AK-47 assault rifle slung on his back. "You want to talk to him?"

"Fuck it," Spann responded.

Spann, dressed in blue jeans and a black sweater, was playing the role of the interrogator while Tyson would photograph each of the enemy fighters as they were brought out. Talking to a Taliban prisoner who would later be identified as 20-year-old American John Walker Lindh, Spann laid out the man's bleak options. Not knowing Lindh was a U.S. citizen, the CIA agent told him that if he didn't start providing information he would be left to die in the prison.

What exactly happened in the courtyard that would lead to Spann's death is still not clear. But at some point, the Taliban were able to overpower the small number of Northern Alliance guards accompanying Spann and Tyson and seize their weapons.

Some reports indicate that a Taliban armed with a grenade ran up to Spann and in a deadly embrace detonated the explosive, killing both men. Another version is that Spann was captured alive and tortured for several hours before the Taliban pumped two bullets into the back of his head.

Some eyewitnesses interviewed by Arnim Stauth and his German TV crew recalled that Spann was in the middle of questioning a prisoner when several Taliban jumped on him. Spann managed to grab his 9 mm pistol and shoot one of his assailants while Tyson killed another. The last image that some of the Northern Alliance guards had of Spann was of the American being swarmed by a large group of prisoners. **

Whatever the sequence of events, the Taliban were now armed with AK-47s

---

**There have been several reports in the British press that a Special Boat Service NCO was awarded a Medal of Honor for his role in saving Dave Tyson. The award and the details of what the SBS operator and his team did to merit the highest military award the U.S. government has to offer were reportedly sealed by an Act of Congress. Whether these reports are accurate has not been established. Most accounts of how Tyson escaped mention that he did so on his own. None of the TV footage shot when Tyson and Spann were interrogating the prisoners shortly before they were taken from Northern Alliance guards and the battle for Qala-i-Jangi began. Up along the angled walls of the fortress, Dostum's men fired down into the crowd of prisoners. Some Taliban shot back while others untied their comrades and freed those locked up in a nearby building. It didn't take long for the prisoners to take control of an ammunition storage bunker and armory in the fortress. Now the Taliban had access to mortars, rocket-propelled grenades and more AK-47s.

Tyson had managed to escape from the courtyard and make his way to the north end of the fort carrying his pistol and an assault rifle. Sounds of gunfire and explosions could still be heard coming from the south end of Qala-i-Jangi. Dostum had stationed only 100 of his men at the fortress and they were now engaged in a deadly battle with their former prisoners.

Tyson ran into a building where Stauth and his television crew had taken cover. Using the journalist's satellite phone, he made a frantic call to the American Embassy in Tashkent, Uzbekistan. The Northern Alliance still held the north end of the fort, Tyson informed embassy officials, but the south section was fully in the hands of the Taliban.

"There's hundreds of dead here," he said. "I don't know how many Americans were killed. I think one was killed. I'm not sure. I'm not sure."

"We just need help to free this place up," he continued. "We can't hit it from the air."

An emergency call was put out for special operations units in the area to head immediately to Qala-i-Jangi.

Shortly after 2 p.m., four truckloads of British and American SOF arrived at the fort. There were 14 operators in total, most armed with an assortment of M4 carbines, AK-47s and the more modern AK-74. The commandos were able to contact Tyson, still inside the fortress. Spann was missing in action, Tyson told them, while he was trapped with the journalists with almost no ammunition left.

Overhead, U.S. fighters and bombers, already alerted to the uprising, were circling, their white contrails cutting streaks across the clear blue sky. Huddled together, the American operators ran down their options. There were too many Taliban inside for an outright assault, so they decided to use aerial firepower to break the siege.

To conduct the precision bombing within the close confines of the fortress, the

---

attacked show any SOF in the courtyard, although that could simply indicate that any operators present had stayed out of camera range. However, there is at least one glimpse in footage shot the day before of what appears to be a western-looking SOF searching some of the Taliban as they came off trucks in the fortress. As well, British special operations troops were almost immediately on the scene, suggesting that they may have been in Qala-i-Jangi at the time of the uprising.

pilots would need the help of special operations troops under the command of Green Beret Major Mark Mitchell. These soldiers would be needed to mark or "paint" targets with their laser designators so the precision-guided munitions (PGMs) dropped by the aircraft could pinpoint that high-intensity beam. In other cases, the men would radio fighter pilots the geographic co-ordinates of a target so they could punch the data into their fire control systems for the PGMs. To do that job, the SOF teams carried Global Positioning Systems, or GPS, which accurately determined ground co-ordinates by using satellite signals.

Within minutes of the first target being marked at Qala-i-Jangi, an explosion rang out and the Northern Alliance soldiers who had crouched along the walls of the fortress started to cheer. At the north end of the complex, Tyson and the trapped journalists counted the bombs being dropped. There were at least nine. Shortly before nightfall, the CIA agent and reporters were able to make their way to the top of one of the fortress walls, scramble over and slide down the earthen slope on the other side.

The fighting continued the next day with the Taliban directing its gunfire up at the walls of the fort where Northern Alliance and special operations troops were situated, looking down into the courtyard. Northern Alliance soldiers would fire over the wall in a typical Afghan style: few would aim, instead preferring to hoist their AK-47s above their heads, high enough to clear the wall, and then pull the trigger hoping that the spray of bullets was enough to hit one of the enemy.

By Monday, two days into the uprising, the Northern Alliance had established a command post inside the compound and had brought in one of their T-55 tanks to fire at enemy positions. Spann's body was found later that day by a Northern Alliance patrol; it had been booby-trapped with a grenade.

Special operations forces on the scene made a decision to call up more air strikes, this time packing even more firepower. "Roger, be advised at this time we're going on the building that's in the centre of the compound," the radio crackled as the fighter pilot circling overhead began his bomb run. "There's Taliban inside that right now."

Minutes later, a 2,000-pound Joint Direct Attack Munition (JDAM) - a satellite-guided "smart bomb" - struck inside Qala-i-Jangi sending a massive brown-grey plume high into the air. Just outside the compound, a group of Green Berets and Air Force combat controllers watched as the cloud rapidly mushroomed and debris began rolling their way.

"Shrapnel inbound!" one of the SOF yelled as the soldiers hunkered down beside a small culvert on the road. "Shrapnel inbound!"

The cloud of debris rolled over the men and as it cleared the area a Northern

Alliance liaison officer suddenly realized a terrible mistake had been made. "That was wrong, that was absolutely wrong," he shouted.

The Green Berets crowded together, staring at the fortress and talking with their Afghan allies. Nearby, Lieutenant Bradley Maroyka and his men from the U.S. Army's 10th Mountain Division, who had been brought in to help in the battle, watched the chaos in stunned silence. "Oh my God, we may have killed the wrong people," one of the SOF soldiers said.

Instead of targeting the Taliban building, the JDAM had plowed into the Northern Alliance's tank and command centre. The massive explosion tore off the tank's turret, flipping the 36-tonne vehicle onto its back. Northern Alliance soldiers staggered from the fortress bleeding and covered in dust, while others worked to pull out three Green Berets who were buried in the debris. Five Northern Alliance soldiers were killed and five Americans wounded by the JDAM.

It was later discovered that the co-ordinates detailing the location of the Green Berets and Northern Alliance command centre inside the compound, which were provided to prevent just this sort of "friendly fire" mistake, had been incorrectly entered by the fighter pilot into the JDAM's guidance system.

As the Taliban continued their determined resistance, the Americans decided to bring in a new weapon to try to quash the uprising. That night around 11 p.m., the drone of engines from an AC- 130 gunship could be heard overhead and the aircraft, a Hercules transport plane converted into a virtual flying tank, started to rain down fire from its 40 mm cannon. The gunship also pumped rounds from its 105 mm howitzer into the fortress and it wasn't long before several of the buildings were ablaze. Yellow and orange fireballs rose into the sky as Qala-i-Jangi's ammunition dump blew up.

By dawn, after consulting with American special operations troops, the Northern Alliance prepared for what it hoped would be the final assault on the compound. Standing on the walls of the fort, they fired RPGs into the courtyard. Some troops were hit by return Taliban gunfire and tumbled off the wall and down the embankment on the other side.

Inside the south area of the fort, Northern Alliance troops began their advance, fighting building to building as they went. The Taliban furiously clung to each portion of Qala-i-Jangi that they controlled. Some popped out of bunkers and buildings from behind Dostum's soldiers and opened fire on them.

When Northern Alliance troops approached a hole in the ground in the courtyard, they were greeted by a barrage of AK-47 fire from the Taliban hidden inside. A grenade was tossed into the opening and after it exploded the soldiers once again began their advance. Amazingly, they were again fired on by the Taliban still alive in the hole.

Some enemy were hiding out in the basement of one of the buildings, firing through small windows at the advancing Northern Alliance troops. In an effort to flush out those Taliban, Dostum's men threw several five-gallon cans of gasoline, followed by grenades, through one of the basement windows. The concoction produced an orange blaze but the survivors still hung on underground.

Against such dogged determination, the Northern Alliance assault temporarily faltered and its soldiers retreated. They re-entered the fortress later in the day, this time backed by tanks which crushed bodies of dead Taliban as they rumbled along, firing at point-blank range at hold-outs still shooting from bunkers.

More British and U.S. special operations troops began arriving at Qala-i-Jangi to survey the situation from the top of the fortress walls. They appeared frustrated that the Taliban were still hanging on despite the unrelenting attacks and angry that journalists were on hand to record the battle. SOF soldiers tried to block one of the cameras with their hands while another threatened several times to kill Arnim Stauth if he didn't tell his news crew to stop videotaping.

Back in Washington, Pentagon officials described the uprising as a "riot" by the prisoners. It was nothing of the sort. American and British SOF and their Northern Alliance allies were involved in a full-blown battle against a tenacious enemy.

After four days, Qala-i-Jangi showed the effects of the intense fighting. Its once green trees laid splintered on the ground. Spann's pickup truck was a burned out hulk. The carcasses of as many as 50 horses lay rotting in the parade square. Body parts and decapitated corpses were scattered around the courtyard.

By Thursday, a group of Taliban was still holed up in a basement, firing on anyone who approached the building. Since tossing grenades into the hole wasn't forcing the enemy out, Dostum's men decided to flood the room with freezing cold water. Inside, the Taliban struggled to stay alive as the water filled the basement up to their necks and the night temperature dropped below freezing.

By Saturday morning, 86 Taliban, some burned, others suffering from hypothermia, emerged from the cellar. Among those was the American, John Walker Lindh.

Before he was taken into custody by the Green Berets at the fortress, Walker Lindh was videotaped by CNN correspondent Robert Pelton. The American Taliban would later admit being in al-Qaeda training camps on several occasions where he saw Osama bin Laden. He didn't hesitate when he told Pelton that his goal was to be martyred in battle.

"Was this the right cause?" the journalist asked the emaciated and filthy young man. "Was this what you thought it would be?"

"It's exactly what I thought it would be," Walker Lindh replied.

The brutal siege at Qala-i-Jangi would be just one episode near the beginning of a war that was to reach around the world. This global, American-led battle to strike at terrorists wherever they operated would be, much like the fighting at Qala-i-Jangi, unrelenting, dangerous and dirty.

On the front lines of the war on terrorism would be a breed of highly trained and motivated soldiers: the special operations troops of the armies of the West and, in particular, those of the United States. Even before the September 11 attacks, it was becoming increasingly clear that this lethal group - some 47,000 shadow warriors working under the umbrella of the U.S. Special Operations Command, or SOCOM - would no longer be confined to the edges of military operations as it had been in the past.

U.S. Defense Secretary Donald Rumsfeld was already in the process of transforming the American military, shaking up what he considered old-style thinking among some of the country's top generals. The former Navy pilot believed the U.S. armed forces was still set up to battle a Cold War enemy. Its generals were reluctant to deploy their units without a time-consuming buildup of large forces heavily reliant on armor and artillery. Rumsfeld saw his transformed military made up of smaller and more rapidly deployable units that made the most of technology, whether it be smart bombs or space-based sensors.

In early June 2001, Rumsfeld summoned SOCOM's top officers for a private meeting at the Pentagon and outlined his vision of the new American war machine. Air Force Brigadier General Gary Heckman, a senior SOCOM official, walked away from the session sure that SOF were going to be the niche warriors of the American military force of the future.

His assessment was bang on.

In the years following the September 11 attacks, there would be an emphasis on U.S. special operations forces as never seen before. Rumsfeld would name a former Delta Force officer as the new Army Chief of Staff, the first time in history that a senior SOF man had been put at the helm of the army. Commandos would be deployed to hot spots, from Afghanistan and Iraq to Yemen and the Philippines.

Just a month before the Qala-i-Jangi uprising, Rumsfeld laid out his game plan as to how he saw the new global war on terrorism unfolding in the coming years. "The only defense against terrorism is offense," he told a Pentagon briefing. "You simply have to take the battle to them."

Doing that would be the job of SOF.

INSET: Johnny "Mike" Spann, a CIA Special Activities Division officer, was killed during the uprising at Qala-i-Jangi. (PHOTO COURTESY SPANN FAMIILY)
MAIN: U.S. Army Green Berets and Air Force combat controllers man an observation post in northern Afghanistan in the early days of the war. (COURTESY SOCOM)

TOP: The AC-130 gunship saved the lives of countless SOF in Afghanistan by directing precision firepower at enemy forces. The aircraft was used during the uprising at Qala-i-Jangi to destroy the ammunition dump in the fortress. (COURTESY USAF)

LEFT: SOF personnel discovered a propaganda poster, which boasts of Osama bin Laden's role in the September 11 attacks, during a raid on an al-Qaeda complex in Afghanistan. (COURTESY USN)

OPPOSITE PAGE: U.S. Air Force combat controllers train at a forward-deployed location during Operation Enduring Freedom. (COURTESY USAF)

America Strikes Back

2 It was perhaps fitting that the first Americans to enter Afghanistan in the wake of the attacks on New York and Washington were CIA agents. After all, almost two decades earlier the spy agency had, arguably, helped create the very situation in that country that the U.S. now had to deal with.

Two weeks after the September 11 attacks, a CIA helicopter skimmed over the Afghan mountain tops, its pilots scanning the rocky terrain for any sign of Taliban or al-Qaeda forces.

Their Russian-built Mi-17 cruised toward a landing zone in the Panjshir Valley, north of Kabul, where they would rendezvous with senior leaders of the Northern Alliance. The passengers, a 10-man CIA team, examined the gear they were hauling. There were weapons, food, computer laptops and radio communications equipment. The most important piece of cargo, however, was a large metal box containing $3 million in U.S. $100 bills.

The CIA agents had a straightforward mission: buy the United States a proxy army that, with the help of special operations forces scheduled to arrive later, would be capable of overthrowing the Taliban regime and rooting out al-Qaeda.

Ironically, the last time the CIA was on a large-scale mission involving Afghanistan its agents were helping to arm and train the predecessors of the very group it now planned to fight. Throughout the 1980s, the CIA had poured an estimated $3 billion into outfitting and arming the Afghan Mujahedeen for their war against the Soviets. President Ronald Reagan described the tribal warriors as "freedom fighters" and gave the agency almost *carte blanche* to ensure the Afghans had what they needed to defeat Russian troops, who had invaded the country to install a puppet Communist regime.

Members of Pakistan's Inter-Service Intelligence agency ran the program to train the Mujahedeen, but it was American dollars and CIA expertise that fueled it. Former British SAS and SBS operators were brought in by the Americans to teach the Afghans combat skills and some Mujahedeen had even traveled to secret camps set up in Scotland and England for such purposes. British intelligence operatives with MI6 provided details on Soviet troop movements as well as limpet mines so the Mujahedeen could sink Russian ships along the main waterway between the Soviet Union and Afghanistan.

As part of the ISI-CIA campaign against the Russians, an estimated 35,000 Muslims from around the world traveled to Afghanistan to join the Holy War or Jihad. It seemed like a brilliant plan at the time: use the Afghans to create a Russian Vietnam.

The tenacious Mujahedeen didn't disappoint. It would take more than a decade, but the Soviets eventually pulled their troops out of the country, leaving behind 13,000 Russian dead. Another 40,000 Russian soldiers had been wounded and the Soviet military had been shaken to its core by the Afghan debacle.

With its mission accomplished, the CIA pulled the plug on its Afghan operations and the one-time freedom fighters were largely left to their own devices. It was simple politics at play. The U.S. interest in Afghanistan wasn't in nation-building, it was about a proxy war with the Soviets. In 1992, the Communist regime in Afghanistan collapsed when the Soviet Union broke up.

Afghanistan was plunged into anarchy as various factions jockeyed for power. By 1996, one group, the ISI-supported fundamentalist Islamic Taliban regime, controlled most of the country. The Northern Alliance, a loose collection of more than a dozen factions, managed to hold onto a small portion of Afghanistan, mainly, as its name suggests, in the northeast of the country. Meanwhile, the Islamic extremists who had come from Saudi Arabia, Algeria, the Sudan, Egypt and 30 other countries to fight in the war against the Russians, continued to support the Taliban or lend their combat skills to help their Muslim brothers in other wars. Now these foreign Mujahedeen fought the Northern Alliance or turned up on battlefields in Bosnia, Chechnya and Algeria.

In 1996, the CIA found itself once again focused on Afghanistan after the Taliban laid out the welcome mat for Osama bin Laden and his al-Qaeda network. Bin

Laden had bankrolled part of the Afghan war with the Russians and was now doing the same for the Taliban in its fight against the Northern Alliance. His al-Qaeda network provided troops, arms and money to the Taliban. In return, Taliban leader Mullah Mohammed Omar allowed al-Qaeda to operate freely from Afghanistan.

But just as he despised the Russians, bin Laden never concealed his intense hatred for the United States. In 1993, fighters from his al-Qaeda network had reportedly taken part in a deadly ambush of U.S. special operations forces in Somalia and bin Laden had vowed to attack America directly.

In August 1996, the Saudi issued a public fatwa, or religious decree, calling for bombings of Western military targets in the Arabian peninsula. Seventeen months later, he broadened that threat with another fatwa, advising his followers to begin attacks on U.S. civilians and military personnel anywhere in world. It didn't take long after bin Laden's fatwas for al-Qaeda's operatives to carry out a series of attacks. On August 7, 1998, two car bombs destroyed the American embassies in Nairobi, Kenya, and Dar Es Salaam, Tanzania, killing more than 200 people.

U.S. President Bill Clinton responded by ordering almost 70 cruise missiles to be fired at al-Qaeda training camps in Afghanistan. But the attacks were largely ineffective. Bin Laden and his men had expected such a reaction and cleared out of the bases before the missiles hit.

After the 1998 embassy attacks, American special operations units were eager to try to take out bin Laden. Officers at the Joint Special Operations Command at Fort Bragg, North Carolina, came up with a detailed plan to insert Delta Force into Afghanistan to either capture or kill the Saudi. But senior Pentagon officials took one look at the proposed mission and canned it.

Almost a year before the September 11 attacks, another plan was hatched, again to put Delta Force on the ground to assassinate the al-Qaeda leader. Again, Clinton administrators and senior military officials overruled the operation.

After the strikes on the Pentagon and the World Trade Center there would be the usual finger-pointing about which agency was to blame for not going after bin Laden. Everyone agreed that the SOF officers had been keen to launch such an operation and willing to accept the likely heavy casualties that would have come with the mission.

Senior military officials would later claim they lacked the intelligence to launch such a risky gambit and there were too many logistical problems to overcome. The U.S. did not have allies in the area who would be willing to provide a base from which an SOF team could be launched.

Richard Clarke, a top counter-terrorism official in the Clinton administration, later suggested it was the senior military leadership that constantly told the White House it didn't want to take on such a job.

For its part, the CIA complained that the military's intelligence needs were unrealistic. The spy agency's paid informants inside Afghanistan could certainly locate bin Laden for an SOF team, but they didn't have all the answers the Pentagon wanted before it approved sending in such a unit. "The requirements they sent us included items like, which side of the door are the hinges on, do the windows open out or go up and down," recalled a former chief of a special CIA unit created to track bin Laden. "And it's just not the kind of intelligence we can provide on anything resembling a regular basis."

The September 11 attacks ended all such reticence. Two days after the attack, the CIA came up with a plan that seemed to offer the most reasonable prospect of success. Insert CIA teams into Afghanistan, along with U.S. special operations forces, to help organize and support the 20,000-member Northern Alliance. Pay the alliance commanders to fight the war and bribe other Afghan warlords to switch their allegiance from the Taliban to the American-backed forces. This proxy war would eventually cost the U.S. about $70 million, covering payments to various Afghan factions, as well as supplies and equipment for their soldiers.

The decision to use SOF was not exactly welcomed among some of the Pentagon brass. They had traditionally viewed such units with suspicion, if not outright hostility. SOF were seen as a sideshow by such commanders, with the real fighting left up to armor, artillery, large formations of conventional troops and massive airpower.

But with Afghanistan, the Pentagon seemed to have little choice, especially with President Bush wanting fast action on the ground. As with Desert Storm, a buildup of conventional troops would take months. But unlike the Persian Gulf situation, there were few neighboring countries in the Afghanistan region that were willing to provide staging bases for large numbers of U.S. soldiers.

Despite having made the decision to commit special operations forces, the need to secure a logistics base in the area and receive permission from various countries to fly over their territory meant it would be more than a month after September 11 before the first Green Beret A-Team was inserted into Afghanistan.

While the Americans were busy working the diplomatic circuit to pave the way for U.S. troops on the ground, British special operations forces were already moving to staging areas in the region. On September 18, a group of British men, wearing civilian clothes and carrying heavy bags containing what appeared to be military equipment, arrived at the main airport in the Pakistani capital of Islamabad.

When the September 11 attacks took place, fortune would have it that the British SAS was already in a strong position to act quickly. At the time, SAS operators were in Oman, preparing to take part in Exercise Saif Sareea with Omani troops. Other SAS, who had conducted mountain warfare training in Pakistan, had developed good working relations with the country's special forces, a move that reportedly helped smoothed the way for the British to begin initial reconnaissance operations.

The first reports of contact with the enemy came on September 23 when a group of commandos, identified by British government sources as a four-man SAS team, came under fire by the Taliban near Kabul. British government officials would later suggest that the incident was a minor firefight and that the Taliban wildly opened fire at the SAS operators before beating a hasty retreat.

Two weeks later, the U.S. started its war with air strikes. Most of the Taliban's small air force was destroyed on the tarmac. Air defence systems, various command centres and Taliban government buildings were also hit.

Such fixed assets were destroyed in a matter of weeks but it soon became apparent that SOF were needed on the ground to search out more mobile targets. As well, al-Qaeda, figuring that the Americans would launch air strikes, had once again cleared out of their training camps and moved into fortified positions in various mountain locations. To find those sites meant "boots on the ground" were needed.

Pressure was also mounting to show the American public some tangible results besides the antiseptic photographs and videotapes of destroyed Taliban buildings and aircraft that were issued to journalists at daily briefings.

The Pentagon responded in two ways.

One was covert, with the nighttime insertion of an Air Force combat controller and a Green Beret A-Team led by Army Chief Warrant Officer David Diaz. Two MH-53 Pave Low helicopters landed Diaz's A-Team 555, or Triple Nickel as they were nicknamed, near the former Russian airbase at Bagram.

There were a few hairy moments on the nighttime insertion. The two choppers became separated and on the way in one of the MH-53s almost plowed into a mountain slope. Seconds before impact, one of the pilots yelled, "Pull up. Pull, up," as the rock wall suddenly appeared out of the darkness.

Triple Nickel was met by the CIA team, which had arrived about two weeks earlier. After a short period of getting settled in, the Green Berets set out on their mission. The men slipped into the Bagram airfield and climbed to the roof of the air traffic control tower. Laid out before them, among the remnants of destroyed Russian fighter jets and helicopters, was a large Taliban force. The Northern Alliance held one half of the airfield. On the other half, black turbaned fighters of the Taliban milled about in trenches, surrounded by almost 50 tanks, armored personnel carriers and anti-aircraft guns.

Diaz motioned to a mound of earth in the distance and handed a pair of binoculars to the Northern Alliance general accompanying him. The man trained the binoculars on the object, soon realizing that it was an anti-aircraft gun emplacement. That would be the first target, Diaz told the general.

An Air Force combat controller assigned to the team used a laser designator to paint the target. Minutes later a smart bomb from an F/A-18 Hornet fighter jet plowed into the gun emplacement, destroying it completely. Over the next six hours, the team called in one devastating air strike after another. Five hundred-pound and 2,000-pound bombs sent armored vehicles flying into the air and panicked Taliban scurrying in all directions, not sure where the missiles were coming from or who was firing them.

Even as the bombs were raining down at Bagram, a more public display of America's special forces might was taking place near Kandahar.

Some 120 U.S. Army Rangers parachuted into a Taliban airfield, 100 kilometres southwest of the city, jumping from MC-130 Combat Talon aircraft. At the same time, a Delta Force squadron, backed up by more Rangers, hit a compound outside Kandahar where Taliban leader Mullah Omar had a house.

Grainy green video, shot through a night-vision scope and released to the news media, showed the Rangers parachuting and images of troops searching buildings at the airport. In the footage, Rangers can be seen leaving behind posters and photographs of New York firefighters raising the American flag over the ruins of the World Trade Center.

This was Pentagon public relations at its best. General Richard Myers, chairman of the Joint Chiefs of Staff, told journalists that the raid showed that U.S. forces could operate inside Afghanistan without significant interference from the enemy. Resistance, he noted, was light.

"The mission overall was successful," he said. "We accomplished our objectives."

The resistance, however, ended up being anything but light for Delta Force members who raided Mullah Omar's compound, and it was questionable whether the operation accomplished anything but provide a photo op for reporters. Delta Force saw the raid at the compound as a "total goat fuck," investigative journalist Seymour Hersh would later report.

The special operations soldiers didn't believe the brass had any idea as to how to properly use SOF. Most of their anger was directed at the conventional military officers who were planning and running the war, including General Tommy Franks who headed Central Command or CENTCOM.

Instead of a covert insertion near the Mullah's house, as Delta Force had requested, the raid was preceded by a massive attack by AC-130 gunships. Helicopters then landed about 200 Rangers near the compound where they immediately took up positions to cover the main Delta Force group of about 100 operators. Those men quickly fanned out around the Mullah's home while a number entered the building. They never expected to find the Taliban leader in the house but figured that valuable intelligence data might be gathered. Delta, however, could find nothing of importance and orders were given to withdraw.

As the operators were leaving the Mullah's house, the Taliban hit them with force. Well-hidden enemy fighters began firing RPG-7 projectiles at the American commandos. Heavy machinegun fire raked the compound, sending Delta troops diving for cover.

The Americans appeared outgunned and outnumbered. Twelve Delta troopers were wounded in the firefight, three seriously. Some of the Delta operators quickly formed into smaller four-man teams to cover the withdrawal of the main force to Chinook helicopters waiting nearby. At the same time, the AC-130 gunships were ordered in again and laid down fire on Taliban positions.

Another group of Delta operators, whose orders were to slip into the surrounding countryside and act as a covert surveillance team, abandoned its mission and quickly headed out to meet another chopper. As one of the Chinooks was lifting off under enemy fire, its landing gear was torn away when it struck the top of a building.

The Taliban displayed the Chinook's wheel, claiming they had shot down the helicopter. "This commando attack has failed," Mullah Amir Khan Muttaqi, a senior Taliban leader, told Qatar's al-Jazeera TV network.

In the days following the Delta raid, more Green Beret A-teams, accompanied by U.S. Air Force combat controllers, would enter Afghanistan. Teams 534 and 595, made it into the northern part of the country to hook up with Northern Alliance units near Mazar-e-Sharif. Others would soon follow. The Green Berets would be under control of the newly formed Task Force DAGGER.

Another group of Delta Force operators and members of the Naval Special Warfare Development Group (formerly SEAL Team Six) would form Task Force 11 and be given the job of hunting down the Taliban and al-Qaeda leadership.

Now the war would begin in earnest.

Interrogations of Afghan prisoners later revealed that both al-Qaeda and the Taliban expected the Americans to retaliate against the September 11 attacks with large numbers of cruise missiles and then back off. They did not expect the U.S. to put troops on the ground.

Interceptions of radio and satellite telephone communications between Taliban commanders also indicated the regime was not all that impressed with the air strikes in the opening days of the war. The aircraft tended to go for fixed targets such as fuel dumps, air defence systems and Taliban government buildings.

But that was before the appearance of the "death ray," the Northern Alliance's name for the laser designators the Air Force combat controllers used. The laser designators made the early days of the war seem like a turkey shoot. Point the laser, guide in the smart bombs, and the Taliban trenches would disappear in a massive plume of dirt and smoke.

It didn't hurt that Taliban soldiers were undisciplined and poor fighters, with one U.S. general going as far as to describe them as "dumber than a bucket full of rocks."

Rarely did the Taliban try to camouflage their armored vehicles and tanks. Troop emplacements and trenches weren't concealed. As well, the Taliban often set up their positions on the crests of hills where they were silhouetted against the skyline and easily targeted by the Green Berets and combat controllers. Their failure to understand what they were up against made it easy for a military that possessed the most sophisticated technology in the world to take apart the regime bit by bit.

Air Force combat controller Master Sergeant Bart Decker recalled the simplicity of calling down devastating firepower on the Taliban. One day in late November he was positioned on a hill south of Mazar-e-Sharif when he saw Taliban retreating down the highway in a column of Toyota pickup trucks and SUVs. The enemy force did nothing to conceal their presence in the dwindling light and the vehicles drove down the road with their headlights on. Using his Global Positioning System, Decker, a 40-year-old Air Force veteran, quickly determined the coordinates for two separate points on the highway, in front and at the rear of the convoy. He then transmitted those to B-52 bombers and F-16 fighters that had been circling overhead. Decker then sat back and watched the bombs land between the two points on the highway that he had selected, obliterating the convoy.

The targets were so plentiful in the opening weeks of the special operations war that another Air Force combat controller was able to call in as many as 10 to 30 air strikes a day for almost a month straight. With each explosion, the combat controller excitedly transmitted into the radio, "Shack on target," indicating that the bomb hit exactly where it was supposed to.

At times the attacks appeared to create a bizarre mix of ancient and new world warfare. Team 595 watched in amazement as Northern Alliance troops charged in a full gallop cavalry attack on Taliban positions, while Green Beret-directed smart bombs slammed into the enemy trenches just seconds before. The horsemen rode through the billowing dirt clouds caused by the explosions, cutting down any Taliban who had survived the bombing.

Back in Washington, Pentagon officials released edited versions of some of the Green Berets' reports to headquarters. "I am advising a man on how to best employ light infantry and horse cavalry in the attack against Taliban Russian T-55 tanks, Russian armored personnel carriers, mortars, artillery, ZSU anti-aircraft guns and machineguns," a Team 595 member wrote in one dispatch. "I can't recall the U.S. fighting like this since the Gatling gun destroyed Pancho Villa's charges in the Mexican Civil War in the early 19th century."

The psychological war conducted by special operations forces also played up the technology advantage held by the U.S. An EC-130 aircraft, outfitted with high-tech broadcasting and jamming equipment, transmitted demands to the Taliban in both Pashtun and Dari dialects that they surrender immediately.

"You use obsolete and ineffective weaponry," warned one broadcast from the plane, dubbed Commando Solo. "Our bombs are so accurate we can drop them right through your windows. Our infantry is trained for any climate and terrain on earth. United States soldiers fire with superior marksmanship."

Stephen Biddle, an associate research professor at the U.S. Army War College, who conducted an extensive study into how the early days of the Afghan war unfolded, noted that on many occasions the SOF teams easily acquired their targets, often at quite a distance away. At Cobaki on October 22, the Green Berets were able to detect a Taliban observation post more than a kilometre from their position. Six days later at Zard Kammar, combat controllers used their laser designators on Taliban defensive positions clearly visible 1.6 kilometres away.

Even when the Taliban were able to counterattack, as they did on November 18 at Tarin Kowt, they moved forward in the open without any attempt to disperse or provide covering fire, wrote Biddle, who conducted interviews of SOF troops and officers for his study. During that assault, the Taliban simply advanced down a road in a column of vehicles at which point they were decimated by smart bombs called in by special operations soldiers.

In another debacle, Taliban reserves were ordered forward to reinforce defences at Bai Beche. Again the column was caught moving in the open and the American commandos painted the targets and called in air strikes. "Officers who surveyed the scene afterward said it brought to mind the infamous 'Highway of Death' leading out of Kuwait City in the 1991 Persian Gulf War," Biddle wrote.

The Taliban also didn't seem to understand the need for communications security. In one incident, a Northern Alliance soldier, pretending to be a comrade, radioed a Taliban commander to ask if he survived the latest onslaught of American bombs. The reply was that he had since the bombs hit 200 metres to the south. The Green Berets immediately corrected their targeting and wiped out the position. Other times the Taliban would talk among themselves on radio frequencies monitored by SOF, giving details as to how much the bombing had missed a particular target. Again, the teams simply adjusted their fire.

On the ground, the Americans were delighted with the results. In one attack, the Triple Nickel Air Force combat controller who went by the name JT directed 45 bombs down on a 300-by-100 metre area occupied by Taliban soldiers. "It was beautiful," he told the *Washington Post*. "The whole area was laden with machine guns and mortars. We completely smoked everything."

The real danger to U.S. commandos seemed to come from their own forces. During fighting outside Kandahar on the morning of December 5, an Air Force combat controller miscalculated on the GPS co-ordinates and called in a 2,000-pound bomb on friendly forces. It was later revealed that instead of transmitting the coordinates of the enemy, he had relayed the location of a Green Beret A-Team to the B-52 bomber overhead. Three Green Berets were killed, another four wounded. At least 20 Afghan soldiers died or were wounded. Hamid Karzai, who earlier that day had been named the country's interim leader, narrowly missed being killed in the incident. He was in a nearby schoolhouse and was slightly injured when hit in the face by shrapnel and glass.

Even with such unnecessary and tragic friendly fire incidents, it didn't take long before the Taliban regime started crumbling under the combined might of advancing Afghan troops loyal to the U.S. and the special operations units that advised them. On November 9, the key northern city of Mazar-e-Sharif fell. Four days later, anti-Taliban forces entered Kabul. The limited U.S. force of a little more than 100 CIA officers and 316 special operations soldiers combined with massive firepower and high-technology targeting had all but brought the radical Islamic regime to its knees.

The experiences of Team 595 were typical. Assigned to accompany Northern Alliance General Dostum's men, it helped liberate six northern provinces and 50 towns and cities. The team destroyed hundreds of Taliban vehicles and killed an estimated 2,000 soldiers, most the victims of smart bombs. For its part, Triple Nickel was credited with killing as many as 3,500 Taliban and al-Qaeda troops and destroying up to 450 vehicles.

The progress of the war was helped immensely by the fact that military planners had been able to significantly reduce the time needed to bring in aircraft to
hit their targets. Previously it had taken three days of planning and relaying targeting information before a location was hit. That was chopped down to 12 hours
in the opening days of the Afghan war. Even faster response times were gained
when planners switched to a system where Afghanistan was divided into 30 "kill
boxes." Aircraft simply orbited each box waiting to be called in by special operations teams to hit targets.

Thanks to this type of efficient targeting and to the highly accurate smart bombs,
Triple Nickel had successfully broken the Taliban's hold on the Bagram airbase.
However, the enemy had managed to retreat to the hills surrounding the former
Russian installation and therefore still posed a threat.

The U.S., which needed the airfield operational so both humanitarian and military supplies could be brought in, turned to the British Special Boat Service. On
November 15, a group of 100 SBS commandos flew into the base on board a Royal
Air Force C-130 Hercules and soon were setting up portable air traffic control and
radar systems. Observation posts were established on the perimeter of the airfield
and minefields were marked. Unexploded ordnance, left over from decades of
war, was cleared.

Just as important were the covert SBS patrols into the hills around the airbase
to locate pockets of Taliban resistance. Once those were discovered, air strikes
were called in. In a matter of days, the Bagram airfield was back in operation with
the first flights bringing in U.S. special operations forces and CIA.

Almost 500 kilometres to the south, the British SAS was hunting down al-Qaeda.
In particular, one battle for a cave complex outside Kandahar in late November
would go down as a defining moment of the SAS role in the war, and prompt a
recommendation in England that one of the operators be awarded the Victoria
Cross for his heroism.

As many as 60 SAS took part in the attack, some engaging in hand-to-hand
combat with al-Qaeda fighters in the dark, cramped confines of a maze of tunnels.

The raid started after an SAS reconnaissance team located the entrance of the
cave complex being guarded by two al-Qaeda sentries. The British commandos
also found another tunnel leading from the underground base that would likely
be used by the enemy as an exit. The unit positioned men at that location while
others crept into position for an attack on the main entrance.

Once everyone was in position, the troopers heard the squadron leader give
his order for the attack to begin. SAS snipers hidden in the rocks killed the two
sentries while other troopers dropped white phosphorous grenades and high ex-

plosives down ventilation shafts they had discovered. As other SAS operators began their advance, al-Qaeda inside the cave opened fire. Two SAS men were immediately wounded in the firefight, the most serious being a 26-year-old trooper who was hit in the abdomen, leg and arm.

Once inside the tunnel, the fighting turned particularly dangerous because of the close confines and the problem of ricochets from AK-47 rounds off the rock walls. The al-Qaeda retreated deeper into the cave system as the SAS continued to fire and toss grenades as they moved forward. As they pressed on past al-Qaeda bodies slumped up against the rock walls, the SAS operators pumped several rounds into the enemy corpses to ensure the men were indeed dead.

The fighting continued for four hours with 27 al-Qaeda being killed. Another 30 were wounded and as many were uninjured but taken captive. Two more SAS men had been wounded while fighting inside the cave.

The prisoners were handcuffed with plastic restraints and airlifted to an inter-rogation centre by U.S. helicopters. The four wounded SAS soldiers were flown back to Britain for treatment at the Centre for Defense Medicine in Birmingham.

One of the wounded men, an SAS regimental sergeant major, had been shot twice as he led his men in the attack. At one point he and his troopers had fought hand-to-hand with al-Qaeda, prompting calls for the soldier to be recognized with a Victoria Cross. Although that wouldn't be the case, the actions of the SAS that day earned various other medals including Military Crosses. The awards were later presented to six SAS men in a secret ceremony in England.

Other Delta and SAS teams concentrated on hitting Taliban supply and fuel lines. In late November, there were at least a dozen attacks on fuel trucks heading from Pakistan or Iran into Afghanistan.

A typical operation took place at night in the middle of the desert near Tungi village, 16 kilometres from the Pakistani border. Two truck drivers, who pulled off the road to sleep, were resting beside their vehicles when a six-man special operations team crept up and put guns to their heads.

"Who are you?" one of the commandos demanded in poorly spoken Persian.

The drivers claimed they were taking fuel from Iran into Afghanistan for farmers but the operators weren't buying that explanation. The soldiers told the drivers that their trucks belonged to terrorists and would be destroyed. As the two men were handcuffed, the commandos used their radios to call for extra firepower. A helicopter soon appeared and blasted the trucks with rockets. The fireballs from the exploding fuel tankers could be seen for kilometres and the special operations team slipped back into the darkness leaving the two drivers unharmed.

With Taliban forces on the run and pro-American Afghan fighters closing in on the regime's stronghold of Kandahar, the focus for U.S. special operations forces shifted to a four-square-kilometre mountainous area with the rather mystical-sounding name of Tora Bora. Located about 60 kilometres southwest of Jalabad, Tora Bora had been a refuge for the Mujahedeen during their war with the Russians.

U.S. intelligence had intercepted radio and satellite phone communications indicating that at least 1,000 al-Qaeda and foreign Taliban had headed to the mountain enclave near the Pakistani border. There were also intriguing snippets of intercepted conversations that suggested bin Laden himself was with the group, hiding out in one of the many caves that dotted the forested ravines.

Because of the limited number of U.S. troops then in Afghanistan, the CIA and SOF had to cut a deal with three anti-Taliban warlords to supply fighters for the operation. More than $1 million, in fresh $100 U.S. bills, reportedly changed hands as the CIA arranged with warlords for an army of 2,500 Afghans to move into the area.

Heavy bombing began on November 30 as B-52s and F-18 fighter jets plastered the area with 500- and 2,000-pound bombs. More than 150 Afghans, who lived near the front-line, were killed as the massive bombing campaign flattened their villages, along with al-Qaeda tunnels.

As the Afghan troops moved into the two valleys believed to hold most of the foreign fighters - many from Algeria, Yemen and Saudi Arabia - they met with heavy resistance. From their cave hideouts, the al-Qaeda fired mortars and RPG rounds. Other times they waited until Afghan forces moved up the mountain and then ambushed them from well-protected positions among the rocky slopes. The American-hired troops were quickly finding out that this type of fighting wasn't the same as having Air Force combat controllers wipe out poorly concealed Taliban fighters with smart bombs.

The al-Qaeda had already shown they were disciplined and effective soldiers. Unlike the Taliban, they used cover and maneuver tactics. In the battle along Highway 4, south of Kandahar, al-Qaeda fighters had patiently hidden in culverts and burned out tanks along the road before ambushing U.S. and Afghan troops.

In response to the resistance at Tora Bora, the Afghans brought up three Soviet-made T-55 tanks which they began to fire into the enemy hilltop positions. The U.S. stepped up its air strikes, while Delta Force and the SAS moved into the valley to begin their own hunt for bin Laden. SOF snipers were perched among the rocks, occasionally killing al-Qaeda who had made the mistake of breaking cover.

By December 10, the Afghans had overrun a few tunnels and the pro-American warlord Hazrat Ali was predicting Tora Bora would soon be in his hands. As added firepower, the Americans also dropped the first of several 15,000-pound Daisy Cutter BLU-82 bombs on one of the cave complexes, the massive blast collapsing tunnels and shaking the ground for kilometres.

Two days later, however, American forces would learn a hard lesson in working with the warlords. One of the Afghan commanders unilaterally arranged a cease-fire with his al-Qaeda counterpart who claimed 100 men were willing to surrender. Under the terms agreed to by the Afghans, the al-Qaeda would give themselves up on December 12 at 8 a.m. and be turned over to the United Nations. It is now believed the deal was a ruse to allow al-Qaeda fighters to slip over the border during the pause in the fighting. Some have even suggested the Afghans cut a deal with al-Qaeda that in exchange for a hefty payment they would look the other way as they headed across the border to Pakistan.

When the American SOF found out about the cease-fire, they were furious, although it really shouldn't have come as a surprise. In the cutthroat world of Afghan politics and warfare, groups willingly switched allegiances if it suited them or was lucrative enough.

The Green Berets ordered B-52s to lay down a heavy bombardment at 8 a.m., sending a clear signal that there would be no compromises. Two hours later, another Daisy Cutter was dropped in the valley.

On December 14, two U.S. special operations soldiers were wounded in fighting as they attacked a machinegun nest. One was grazed in the shoulder by a round while the other was hit in the knees. Spectre gunships joined in the battle, using their howitzers to blast apart the machinegun nest hidden in a tunnel.

Back in Washington, deputy U.S. Defense Secretary Paul Wolfowitz told journalists that intelligence reports indicated bin Laden was indeed in the Tora Bora area. "This is a man on the run, a man with a big price on his head," Wolfowitz said at a Pentagon briefing. "He doesn't have a lot of good options,"

But bin Laden did have an option; a 16-hour trek would take him to the safety of the Northwest Frontier in Pakistan. Although Pakistani troops were supposed to have sealed the border, it was a difficult task, especially in the lawless areas of the frontier where Pakistan had little power over various tribes there.

By the end of the fighting, at least 400 of bin Laden's men had been killed at Tora Bora. Some 300 bodies were found scattered throughout the two valleys while fresh graves for another 100 were also later discovered. But it is estimated that more than 700 al-Qaeda, most of the main force trapped at Tora Bora, were able to cross over the border to safety.

It wouldn't be the first time that the enemy had made an escape. Several weeks before the Tora Bora operation, the Northern Alliance, along with U.S. special operations forces, had surrounded the Taliban and al-Qaeda in the town of Kunduz. Trapped with them, according to reports, were Pakistani Army officers, intelligence advisers, and volunteers who were fighting alongside the Taliban. Since the early 1990s, Pakistan had supported the Taliban regime and, in turn, had a significant amount of influence in Afghan affairs.

On November 23, two aircraft flew into Kunduz in the middle of the night, sparking anger among Northern Alliance commanders who accused the U.S. government of giving their approval to a mass exodus of enemy fighters. American intelligence officials would later suggest the airlift had indeed been organized for Pakistani officers working with the Taliban. The U.S. had given its approval as a way to save face for Pakistani President Pervez Musharraf who had been providing support for the American war. Neither he nor the American government could afford the political embarrassment of large numbers of Pakistani intelligence officers being captured with the Taliban. Unfortunately, an unknown number of Taliban and al-Qaeda fighters and their leaders were flown out on the aircraft as well.

Tora Bora, however, was an altogether different matter. Kunduz may have been the result of the hard reality of keeping Pakistan on side in the war on terror. But how did the U.S. let the architect of the September 11 attacks slip through its fingers?

There are various theories. One is that Osama bin Laden left the area around November 27, escaping as U.S. special operations officers tried in vain to convince their Afghan allies to immediately launch an attack on Tora Bora.

Some CIA agents contend bin Laden was still in the area until mid-December and that U.S. intelligence had intercepted at least one message from Tora Bora indicating the al-Qaeda leader was still directing combat operations in the valleys.

In his study on the Afghan conflict, U.S. Army War College research professor Biddle concludes that the massive American bombing at Tora Bora was simply not enough to compensate for the reluctance among the Afghan proxy troops to fight it out with al-Qaeda forces. "Many now see this ground force hesitancy as having allowed Osama bin Laden and much of his command structure to escape capture and flee into neighboring Pakistan," Biddle wrote in his report.

CENTCOM commander General Tommy Franks would later acknowledge to a Senate Armed Services committee that he wasn't happy with having to use Afghan troops at Tora Bora.

Those in the British SAS blame the timidity of American generals for bin Lad-

en's escape. The SAS believed they had the al-Qaeda leader trapped at one point in one of the valleys and requested permission from U.S. commanders to divide into two groups: one to push through the valley, the other to capture bin Laden on the other side as he tried to escape. Sixty-four SAS, a full squadron, was in Tora Bora and had linked up with their Delta Force comrades for the mission, but U.S. officers, worried about the potential for a large number of casualties, reportedly declined to approve the operation.

Through early January 2002, special operations forces and CIA operatives continued to comb through the Tora Bora caves and gather intelligence. It was an extensive job. Instead of the 40 or 50 caves believed to exist in the area, the teams discovered almost 200 tunnels. Inside they found computer discs, passports and identification documents, as well as large amounts of ammunition.

In Washington, military leaders tried to emphasize that the war wasn't about the hunt for bin Laden. It was about destroying the al-Qaeda network. Now the focus was on another larger cave complex at Zhawar Kili, near Khost. Rear Admiral John Stufflebeem, the Pentagon's deputy director of operations, said U.S. forces would root out and destroy al-Qaeda there.

That inland mission, somewhat unexpectedly, would go to special operations warriors more familiar with fighting the enemy on or near the water – the U.S. Navy's SEALs.

PREVIOUS PAGE:
CENTCOM commander
General Tommy Franks
meets with Green Berets
and other U.S. SOF in
Afghanistan. (COURTESY
SOCOM)

ABOVE: A suspected
terrorist, captured by SOF
in Afghanistan, is held
captive at the U.S.
detention centre in
Guantanamo Bay, Cuba.
(COURTESY DOD)

RIGHT: The first Green
Beret teams into
Afghanistan relied on
mules and other pack
animals to transport their
gear through the
country's rugged terrain.
(COURTESY DOD)

TOP: *Master Sergeant Bart Decker, an Air Force combat controller, rides on horseback with the Northern Alliance troops during the early days of the war in Afghanistan. (COURTESY USAF)*

ABOVE: *A U.S. SOF soldier looks for al-Qaeda and Taliban targets in northern Afghanistan. (COURTESY SOCOM)*

OPPOSITE PAGE: *U.S. Navy SEALs brought their desert patrol vehicles out of storage for Operation Enduring Freedom. (COURTESY USN)*

Edge of the Knife – Task Force K-BAR

3 The Predator unmanned aerial vehicle circled quietly, gliding through the deep black Afghan sky, its human targets on the ground completely unaware they were under surveillance.

The group of Afghan men in the country's Paktia province couldn't hear or see the robotic aircraft high above them as it tracked their progress with its infrared camera and transmitted the images back to its ground control station.

At his headquarters just outside Kandahar, U.S. Navy SEAL Captain Robert Harward received a call on his satellite phone from intelligence officials who had been alerted by the Predator's ground control crew. "Mullah K has left the building," Harward was told.

The veteran SEAL officer called his planners together and over the next 30 minutes hashed out the details of a mission to snatch Mullah Khairullah Kahirkhawa, a former Taliban governor now on the run from coalition forces. Kahirkhawa would be an important catch for intelligence officers as he had been close to the Taliban regime's leader, Mullah Omar, as well as having once served as minister of the interior.

Harward's Task Force K-BAR, a multi-national special operations group with responsibility for southern Afghanistan, had been waiting for weeks for the chance to get a fix on Kahirkhawa. Now, a high-tech eye in the sky had put the Taliban official in their sights.

As the Predator continued to transmit real-time images of Kahrikhawa's group, more than 40 SEALs and Danish SOF were climbing aboard an MH-53 helicopter. For added firepower, an Apache attack helicopter would fly cover for the team.

One hour after Harward received the satellite phone call, the Mullah would be in American hands. The SOF strike force had successfully raided a safe house in the village of Mohammad Hasan where it captured the fugitive.

For Harward, the mission, conducted on February 14, 2002, was a testament to the highly effective teamwork that had become a hallmark of the Combined Joint Special Operations Task Force South, better known as Task Force K-BAR.

The task force, named after the U.S. military's K-Bar knife, brought together SEALs (Teams 2, 3 and 8), U.S. Air Force and Army special forces, U.S. Marines, as well as SOF from eight countries. Included among these were 102 Danish operators, 103 from the German Kommando SpezialKraefte, 40 each from Canada's Joint Task Force Two and the New Zealand Special Air Service, 95 from the Australian SAS, and a combined Norwegian force of 78 from the Jaeger Kommando and Marine Jaeger Kommando. One Turkish special forces liaison officer rounded out the task force.

In total, K-BAR had 1,300 special operations troops and another 1,500 support personnel. The task force had responsibility for 19 separate operational areas, its commandos conducting missions in Afghanistan's deserts as well as its mountains. At times, K-BAR's patrols would reach out as far as 400 kilometres from its main base at the Kandahar airfield.

The task force's orders were straightforward: capture al-Qaeda and Taliban and conduct surveillance and raids of suspected enemy compounds, tunnels and caves. During these "sensitive site exploitation" missions, K-BAR's commandos were to grab everything and anything they could that might be of intelligence value.

The SEALs had a long and respected history of covert operations, from the Vietnam War to missions in Panama, Grenada, Kuwait and Bosnia. But K-BAR would mark the first time that a SEAL would command a joint task force whose primary mission was ground operations.

Harward, who had been involved with naval special warfare for almost 20 years, was well-suited to the job. The 45-year-old SEAL, fit and square-jawed, had served on missions in Kuwait and Bosnia. He spoke fluent Farsi, learning the language as a teenager when his family lived in Tehran, Iran, where his father had worked at the U.S. embassy. Harward was also one of the few U.S. officers involved in the Afghanistan mission, dubbed Operation Enduring Freedom, who had actually previously visited the country. During a summer off from studies at the Tehran American High School, he hitchhiked through Afghanistan.

About a month before the September 11 attacks, Harward had been named commander of Naval Special Warfare Group One based in Coronado, near San Diego. NSW One had responsibility for all naval special warfare missions in the Pacific and Southwest Asia and it wouldn't be long before the specialized skills of its operators would be called upon for the Afghanistan war.

It was a chance meeting in mid-November 2001 on the Persian Gulf island state of Bahrain between Harward and Marine Major General James Mattis that ensured the SEALs would have a role on the front-lines. Mattis, commander of Task Force 58, had just been given the job of putting a 1,000-strong Marine force, who were on board the USS Peleliu and the USS Bataan in the Arabian Sea, into Afghanistan. He would need "eyes and ears" on their selected landing zone, an airstrip some 100 kilometres southwest of Kandahar.

The man he had bumped into on a street corner had just the operators for the job. Mattis, who knew Harward from Coronado, asked the SEAL captain what he was doing in Bahrain. The naval officer explained he was trying to figure out a way to get into the war.

'Well, come on, I think you just found one," Mattis responded.

Two weeks later, members from SEAL Team 3 were conducting surveillance on the airstrip the Marines had designated to become their first forward operating base in Afghanistan. The commandos had been inserted at night a fair distance from their objective and had humped their way to some high ground overlooking the airfield. Not knowing what to expect, each SEAL carried more than 50 kilograms of weapons and ammunition, as well as enough food and water to last at least a week.

It was supposed to be a 24-hour mission but ended up dragging on for 96 hours because the Marines were having logistical problems getting their assault force ready. In their "hide," the SEALs only talked in whispers or used hand signals. Food was eaten cold because the chemical packs used to heat rations gave off a distinctive smell. Not wanting to give any clues as to their presence, the men even went so far as to bag their feces.

During the day, the SEALs scanned the airstrip and desert with binoculars but could detect no movement. The white painted buildings appeared empty. When darkness fell they used their night-vision goggles to scope out any movement around the target area. Again nothing, except the odd desert animal scurrying across the terrain. The only action seemed to be down the road towards Kandahar, the spiritual home of the Taliban, where enemy positions were being pounded by U.S. airpower. At night, flashes could be seen on the horizon as the bombs, sounding like muffled thunder, detonated.

Not seeing any activity on the airstrip, the SEALs decided to move carefully from their vantage point and conduct a closer investigation of the target. It was quickly determined that there were no al-Qaeda or Taliban at the site, so the signal was transmitted for the Marines to move in. Several hours later, on November 25, the first wave of six CH-53E helicopters began to arrive from the USS Peleliu in the North Arabian Sea and the Marines began to settle into their new home, which they dubbed Camp Rhino.

After landing at the airfield, a Marine officer thanked the SEALs for their work and the men boarded an aircraft and headed back to their base in Kuwait.

But for Harward and his officers, the work was just beginning. K-BAR was slowly coming together as operators from various nations started arriving in Afghanistan. The task force would eventually set up its base at the Kandahar airfield, abandoned by the Taliban in early December as U.S.-led forces overran the city.

K-BAR would work differently from the Green Berets' Task Force DAGGER. DAGGER's A-Teams, such as Triple Nickel, were essentially combat advisors to Northern Alliance troops. Harward's men, by contrast, wouldn't be leading Afghan indigenous forces into battle. Although they would have interpreters and guides, K-BAR's missions were designed to be mainly covert intelligence-gathering operations or lightening fast "direct action" raids. Not working with proxy troops lessened the chances of a mission being compromised by Afghan spies.

FBI agents accompanied the K-BAR teams to interrogate any detainees taken during the operations. Decisions were usually made on the spot about whom to free and whom to send back to Kandahar for further questioning. The latter, referred to by the commandos as "keepers," would be held at a detention centre at the airfield. After further questioning there, FBI and military intelligence officials would decide whether to transfer the prisoners to Camp X-Ray, an American jail for suspected terrorists that had been set up at Guantanamo Bay, Cuba.

K-BAR operators also had at their disposal high-tech support which had been lacking in the early days of the war. While Toyota and other four-wheel drive pickup trucks were used to transport some special operations forces, the SEALs also made extensive use of their heavily armed DPVs (Desert Patrol Vehicles), which were flown into specific areas by helicopter.

In addition, the Predator UAVs were flying more missions, proving their usefulness in scouting out suspected al-Qaeda sites in advance of any raids. Another valuable piece of equipment for K-BAR was the U.S. Navy's P-3 Orion aircraft, outfitted with a real-time video surveillance system and other sensors. Harward put a SEAL on the P-3 so he could help determine what was happening at a target location and relay that information back to his command post as well as to the operators on the ground before they hit their objective.

For Harward, the surveillance technology on board the Predator and P-3 had its advantages since it reduced the risk to his men. But such high-tech gear also had limitations since it didn't provide the whole intelligence picture of what was happening on the ground. The Predator, for instance, showed one pinpoint image of a specific target but no details as to what was going on around it. There was no way of avoiding putting boots on the ground.

K-BAR's missions, like those of all special operations forces operating in Afghanistan, were a frustrating mix of successes and "dry holes." In one such case, SEAL teams raided a fortress in the Jaji Mountains near the Pakistani border, believing the compound was an al-Qaeda stronghold. Eighty people in the fortress were rounded up but it was later determined none were al-Qaeda.

In early January 2002, a K-BAR team began a five-day reconnaissance mission on another compound, this time at the mountain village of Shkin, seven kilometres from the Pakistani border. The team had set up its surveillance post on a cliff overlooking the village. The operators spent a miserable several days, shivering in freezing temperatures and wet snow, as they covertly tried to determine if the site was an enemy compound.

Two days into the mission, one of the men developed kidney stones, but instead of compromising their position by radioing for a helicopter to take the operator out, the team decided to carry on. The ill man was bundled up in a sleeping bag and given painkillers as his comrades continued with their surveillance.

After several days, the reconnaissance team radioed for other K-BAR operators to make their move. Helicopters started landing Marines and special operations troops who swept into the compound, taking into custody 30 people who were found sleeping on the dirt floor of a four-bedroom mud-brick house. The men were interrogated by two FBI agents who suspected that seven may have al-Qaeda or Taliban ties. Those suspects were transported back to Kandahar. Weapons and explosives were also found during the hour-long mission and destroyed by K-BAR commandos.

Not all operations, however, went so quickly. In early January, American commanders decided to strike at an al-Qaeda complex near Zhawar Kili after intelligence officials had reported "some activity" in the area.

K-BAR officers figured their ground mission at Zhawar, 50 kilometres southwest of Khost, would last from about seven to 12 hours. Instead it dragged on for eight days after the initial SEAL search team found more than 70 caves and tunnels scattered throughout a twisting ravine. Among K-BAR officers, the mission quickly became known as the "Gilligan's Island" operation, a name inspired by the popular 1960s TV comedy in which a group of people go out for three-hour boat tour and end up stranded on an island for years.

The number of tunnels at Zhawar shouldn't have come as a surprise since the complex had a long history of serving as an Afghan stronghold. During their war against the Russians, the Mujahedeen had used money provided by the U.S. government to turn the site into a command and control center and logistics base. Bulldozers had been brought in and explosives were used to carve long tunnels into the side of the five-kilometre-long ravine.

The underground installation was equipped with rudimentary classrooms, living quarters, a first aid station, storage areas and even a mosque. Some of the reinforced tunnels ran for hundreds of metres into the side of the ravine and electrical power was provided by a portable generator.

The Zhawar stronghold had survived several Russian assaults in the 1980s, but after sustaining aerial bombing and artillery barrages, the complex fell to Soviet and allied Afghan commandos in the spring of 1986. Thirteen Soviet helicopters and other aircraft were shot down in the battle and 100 Afghan government soldiers were killed in taking the area. After occupying the site for two days and destroying the ammunition and weapons they found in the tunnels, the commandos withdrew and Zhawar was reoccupied by the Mujahedeen.

Years later, al-Qaeda occupied Zhawar and, like the Mujahedeen, used the installation as a command and control center and training base. The U.S. government had already targeted Zhawar in 1998 when, as a response to al-Qaeda's bombing of U.S. embassies in Kenya and Tanzania, it ordered Tomahawk cruise missiles to be launched at the base. The barrage, however, did little damage as al-Qaeda, which expected such a move, ordered its forces to temporarily vacate the site.

The K-BAR plan for Zhawar called for several days of attacks by B-52 and B-1 bombers, F-18 fighters and AC-130 gunships before commandos were inserted by helicopter just before dawn on January 6. At first, there was a 25-member special operations team and a 50-man blocking unit provided by the Marines to kill or capture any al-Qaeda trying to flee the area. Since the SEALs had initially thought the mission would be a quick strike, they packed light, dumping any extra food and water as well as cold weather clothing.

As the K-BAR operators carefully approached the Zhawar caves, they spotted a group of men near a tunnel about half a kilometre away. Air strikes were called in and al-Qaeda quickly retreated into the surrounding mountains. Their move to the high ground signalled the beginning of eight days of sporadic firefights between the two groups.

Back in Kandahar, it was decided that the strike force would spend the night and continue searching the caves the next day. That evening, the SEALs could see the enemy using blinking lights high up the rocky slopes and although they couldn't understand the code being used, it was obvious the flashes were some kind of communication between the al-Qaeda groups.

On the second day of the mission, it began snowing and the temperature dropped well below freezing. With the team camped at around 2,400 metres above sea level, there were undoubtedly some regrets about the decision not to bring cold weather clothing and extra food and water. As the weather worsened, the operators headed towards a series of abandoned villages near the cave complex where they were able to find extra clothes and blankets. One of the officers would later recall that the situation was so bad several commandos came close to becoming hypothermic.

By day three, word came from headquarters that a more extensive search of the underground complex was needed. Helicopters had resupplied the team with food and clothing and more SOF were being sent in to help scour the maze of tunnels and caves.

As was becoming increasingly common, the SEALs discovered "terrorist escape kits" buried around Zhawar. Each kit contained everything an al-Qaeda operative would need if he had to flee an area. There was a set of fresh clothing, money, a Pakistani passport, an AK-47, a pistol and grenades.

Once inside the caves, the K-BAR commandos discovered just how extensive the al-Qaeda operations at Zhawar had been. There were armored personnel carriers and anti-aircraft guns as well as stacks of artillery rounds and landmines. Piled up in several of the underground rooms were tens of thousands of rounds of ammunition and boxes of explosives. More than half of the 70 tunnels had been reinforced with steel beams.

As the SEALs carefully moved through the labyrinth of dark tunnels, weapons at ready, they came upon a classroom filled with al-Qaeda propaganda. There was a poster of Osama bin Laden superimposed on aircraft crashing into the World Trade Center. Another poster had a photograph of President Bush with blood running down his face. On a wall in one of the classrooms, there were detailed diagrams of anti-tank mines for students to study.

The caves were so long, some running several hundred metres, and so dark that the SEALs couldn't see their hands in front of their faces. They switched on the flashlights located on their M4 carbines and continued. In one of the tunnels, an operator found a child's foot still in its shoe. It was a gruesome reminder that al-Qaeda often lived in the training camps with their families; it was likely the child had been injured or killed during the bombing raids before the K-BAR team landed.

Since the al-Qaeda camp was so vast, the SEALs had two of their desert patrol vehicles airlifted in by helicopter so they could use the DPVs to search the area and move up and down the ravine. When the Afghan war broke out, the heavily armed dune buggies had been sitting in storage in Virginia for six years and had not been well looked after. Before being sent overseas in late December, they were refurbished by Chenowth Racing Products, the company that originally manufactured them. New tires were put on the vehicles, modern radios installed, parts replaced and weapons upgraded.

After the underground complex was stripped of intelligence documents, the SEALs set up charges to detonate the al-Qaeda ammunition stocks. But with the massive amount of enemy ammunition, tanks, trucks and other vehicles, the K-BAR operators didn't have enough explosives to do the entire job.

It was decided that the best way to destroy the complex was to call in air strikes on the caves and tunnels. Zhawar, however, wouldn't go easy; it would take an incredible 180,000 kilograms of bombs dropped from B-52s, B-1s and F-18s to destroy the installation. Huge fireballs filled the sky as aircraft made direct hits on the tunnels. Explosions, caused by a series of chain reactions as ammunition caches deep inside the mountain detonated, could be heard for days.

By the time the mission ended on January 14, the combined force of SEALs, Air Force special operations troops and Marines had killed 12 of the enemy and had taken another eight suspected al-Qaeda operatives prisoner. The number of SOF used in the mission ranged from 75 on the first day to about 200 for the more extensive searches of the caves.

K-BAR continued its sensitive site exploitation with another raid on a complex at Yaya Kehyl in the Paktika province. That mission yielded a treasure trove of equipment and intelligence information, in particular computers containing information about al-Qaeda as well as satellite telephones used by the terrorists. Also seized were the usual stocks of AK-47s and anti-aircraft guns, as well as a kilogram of unrefined opium.

While the raids were successfully turning up large amounts of weapons, the task force was running into problems trying to determine friend from foe. As in most guerrilla wars, the enemy easily melted into the general population. K-BAR used information gathered from intelligence agencies which intercepted radio and satellite communications, as well as tips from informers, but it was far from a perfect system. "It's not a really clear picture but maybe you get a silhouette," K-BAR's intelligence chief, Lieutenant Colonel Kevin Wooton, acknowledged in an interview with *U.S. News and World Report* magazine.

The other concern was the potential of being misled by informants and getting drawn into tribal rivalries. Some Afghan leaders had tried to manipulate special operations troops, informing them that certain villages or groups were involved with al-Qaeda or the Taliban when in reality they were not. While in Kabul, members of the Green Beret A-Team 555 were approached on several occasions by their Afghan comrades with co-ordinates of "enemy" positions they wanted destroyed by air strikes. The requests were refused when it turned out that the targets were a rival faction who were also American allies.

This type of manipulation is believed to have been behind one of K-BAR's most controversial raids which left at least 16 anti-Taliban Afghans dead.

Ten days after the Zhawar operation, two Green Beret teams prepared their gear and weapons for a night-time raid on two compounds at a village near Hazar Qadam, 95 kilometres north of Kandahar. Intelligence information indicated the buildings, nestled in the mountain village of about 2,000 people, were being used by the enemy.

In the early morning darkness of January 24, one group of Green Berets slipped into the village and headed for one of the buildings, a school they believed was being used as an al-Qaeda ammunition dump. In fact, Afghans loyal to the country's interim president, Hamid Karzai, had been operating from the compound, using it as a depot for weapons they collected from former Taliban. The Taliban had left the village months earlier and al-Qaeda had never even been in the area.

The Afghans at the school had indeed been quite effective allies for the new government, having collected 400 mortar bombs, 300 RPG projectiles and more than half a million rounds of small arms ammunition.

Inside the school, more than a dozen men were sleeping on the cement floor while one stood guard. At about 2 a.m., the man on guard heard a noise outside and went to investigate. Peering into the darkness, he saw the reflection of the night-vision goggles worn by one of the Green Berets and the man ran to another building to awaken a senior officer. "The Americans are here," he told the puzzled Afghan who got up to investigate.

Just then, gunfire erupted. The Americans claim they were fired on first and one of the Green Berets was wounded in the ankle. Some Afghans in Hazar Qadam vehemently denied that, saying the men sleeping in the school, while armed, didn't fire a shot. At least one other Afghan at the scene said a man who had the sole AK-47 in the schoolhouse did open fire.

Whatever the case, by the time the gunfire and explosions died down, at least 12 Afghans lay dead on the school floor. Two more bodies were found outside the school. As the Green Berets withdrew from the area, they called in an AC-130 gunship to target the building, peppering it with cannon and mini-gun fire.

At a second target about a kilometre and a half away from the school, the K-BAR team killed two and captured 27 Afghans. Some of those, sent back for interrogation to Kandahar, would later allege they were beaten by the Green Berets. Before the team left, one of its men stuck a "God Bless America" sign on a pickup truck destroyed in the raid. Also on the sign someone had written, "Have a nice day. From Damage, Inc."

After the raid, there were angry accusations against the K-BAR operators. Afghan villagers say they found two of the dead men at the school with their hands tied behind their backs with white plastic straps, the allegation being that they were executed while restrained. Others alleged they heard the men in the school trying to surrender, yelling, "We're friends, we're friends," in Pashtun, as the Green Berets continued shooting. The villagers also said the number of dead was more than the 16 the Americans acknowledged killing. They claimed there were almost 40 people killed in the raid, some of those civilians who lived nearby.

K-BAR commander Harward denied the execution allegations. His men, he said, had little alternative but to open fire when they were shot at.

To further confuse matters, Jan Mohammed Khan, the governor of Oruzgan province where Hazar Qadam is located, said that U.S. special forces who were assigned to him to advise on security matters were told about the weapons collection program at the school.

At first, Pentagon officials denied that the dead Afghans were anything but al-Qaeda. Journalists were told by military sources that the firefight at the school was vicious and had involved hand-to-hand fighting, a clear example that al-Qaeda operatives were fanatics willing to die for their cause.

Defense Secretary Rumsfeld confidently stated that intelligence information that had been gathered in the weeks preceding the mission was solid proof that enemy forces were operating in Hazar Qadam. Not only had there been intercepts of communications to indicate al-Qaeda were in the village but special operations forces had kept the site under surveillance for several weeks, he suggested.

Still, Rumsfeld did acknowledge that the situation on the ground for SOF units was becoming increasingly difficult. "It is not a neat situation where all the good guys are here, and all the bad guys are there," he explained.

Faced with growing publicity over the incident and statements by Hamid Karzai that U.S. special forces had made a major blunder by killing allied forces, the Pentagon launched an internal review of the mission. The investigation concluded that those inside the school and the other compound were not al-Qaeda.

But even with that outcome, Rumsfeld defended the Green Berets, noting that they had little choice but to open fire when they were fired upon. "Clearly, in retrospect, that's unfortunate," Rumsfeld said of the Afghan deaths. "On the other hand, one cannot fault the people who fired back in self-defense."

The Afghans, Rumsfeld stressed, fired first. "Let's not call them, 'innocents,'" he lectured journalists at a Pentagon briefing. "We don't know quite what they were. They were people who fired on our forces."

The 27 prisoners taken in the raid were released and the families of the dead paid about $1,000 each in compensation. American troops apologized to some of the captives as they were being released, giving them new socks, gloves and shoes before sending them on their way.

The Chairman of the Joint Chiefs of Staff, Air Force General Richard Myers, and Rear Admiral John Stufflebeem, deputy director of operations, said there was no proof that the Green Berets beat the prisoners, although the naval officer readily acknowledged being captured by special operations forces was not a happy experience. "And in that initial encounter, you don't know who's good, you don't know who's bad, and you don't take chances," he explained. "So everybody's treated the same, and it's relatively harsh, I would say."

"Once identities are established, it's quite a different mode," Stufflebeem added.

Myers explained that the plastic white straps found on some of the bodies were a standard special operations procedure; people in the target area were subdued until their identities could be determined. It was perhaps possible, according to the general, that some of the Afghans died with the plastic straps on while they waited for medical attention. But he dismissed outright the allegation that the men were executed.

The mission at Hazar Qadam was a tragic mistake but one K-BAR's commandos seemed to take some lessons from. In February, a team was inserted near the village of Ali Kheyle, an extremely remote site some 4,200 metres above sea level.

A Predator had transmitted images of about 50 armed men moving in and out of the fortress-like compound near the city of Khost in eastern Afghanistan. That in itself didn't mean much as almost every male in the country carried an AK-47, with some packing heavier weaponry such as RPGs to defend against bandits and rival tribes.

Before moving in on the village, K-BAR planners decided to put a reconnaissance team near the site to conduct surveillance for several days. The high-walled compound tucked in the mountains reminded some of the commandos of something out of an Indiana Jones movie. Even after several days of surveillance, it was too difficult to determine the situation in the village, so K-BAR commanders decided to confront the occupants. With an AC-130 gunship circling overhead, the special operations soldiers used a loudspeaker to demand that those inside exit immediately without their weapons.

K-BAR's caution was worth the gamble. Out of the fortress walked not only the unarmed men but 56 children, along with dozens of women, who had been staying indoors, and thus out of sight, for several days because of the extreme cold. Although the compound at one time had been a Taliban training base, it had long been abandoned by supporters of that regime. Weapons the Taliban had put in storage were promptly turned over to the special operations troops.

The next major mission K-BAR was assigned would mark a shift in how the war was being fought. Operation Anaconda, for the first time in the Afghan conflict, would see special operations troops being used in support of conventional forces, which until that point had been left out of the fighting.

Anaconda would involve troops from the U.S. Army's 101st Airborne and 10th Mountain Division, as well as K-BAR's SEALs and its coalition SOF units, Canada's Joint Task Force Two, the Australian SAS, Norwegian and Danish operators and the German KSK. In total there would be about 200 special operations soldiers from the foreign contingents. K-BAR, as part of the Joint Special Operations Task Force, was made available to Anaconda commander U.S. Army Major General Franklin Hagenbeck, along with Green Berets from Task Force DAGGER. Hagenbeck would also have access to a "black" special operations group made up of British SAS, the U.S. Navy's DevGru (formerly SEAL Team Six) and Delta Force. That group, however, would report directly to CENTCOM commander General Tommy Franks. Also on the battlefield would be CIA special operations forces.

Anaconda, named after the snake that tightens its coils around its prey until the victim dies, was supposed to be a knockout blow against the remaining Taliban and al-Qaeda forces.

The target was the Shah-e-Kot Valley, a mountain region 150 kilometres south of Kabul. Almost two decades earlier, the Afghans had used the valley, dotted with caves and tunnels, as a stronghold from which they had successfully fought off two Soviet assaults.

U.S. intercepts of radio and satellite phone messages, as well as aerial surveillance from UAVs and satellites, had determined that al-Qaeda forces were regrouping at Shah-e-Kot. The best guess put forward by the intelligence teams was that the terrorists were resting in the valley before making their final push over the mountains to the relative safety of Pakistan's Northwest Frontier.

ABOVE: *SEALs with Task Force K-BAR on patrol somewhere in Afghanistan.* (COURTESY USN)

LEFT: *A SEAL stands in one of the Zhawar tunnels, his arms outstretched to show the width of the passageway.* (COURTESY USN)

TOP: *SEALs check out a village during a patrol in Afghanistan.* (COURTESY USN)

LEFT: *A U.S. SOF discovers tunnels crammed with explosives and ammunition during one of Task Force K-Bar's sensitive site exploitation missions.* (COURTESY USN)

Estimates of the size of the group varied widely but it was believed, at least initially, that there were a few hundred enemy fighters gathered in the area. The assault force would later find out differently - it would meet resistance from at least 500 al-Qaeda with some estimates putting the figure as high as 1,000.

Hagenbeck, who commanded Anaconda from his headquarters at Bagram airbase, gave key roles to the special operations forces assigned to the mission. K-BAR commandos would be inserted in advance to conduct reconnaissance and gather more detailed intelligence about the area and the enemy forces that had assembled there.

The Green Berets would lead a main contingent of Afghans, under General Zia Lodin, who were hired and trained by the Americans over a four-week period before the mission. In a classic hammer and anvil operation, the Green Berets and Lodin's troops would drive towards three villages in the Shah-e-Kot valley, pushing the enemy through the mountains where they would be destroyed on the other side by the 101st Airborne and 10th Mountain Division. K-BAR operators would also block off any escape routes to Pakistan through the high mountain passes.

Hagenbeck optimistically saw the operation taking about 72 hours to complete.

The preparations for the battle and how it was fought would be hotly debated in military circles long after Anaconda was over. Part of the criticism focused on the use of Afghan proxy troops provided by General Lodin, a less than reputable commander. It is now believed that at least some of his men were former Taliban who warned their al-Qaeda comrades in advance that the attack was coming.

The other issue still under contention was the decision not to start Operation Anaconda with a massive aerial bombardment. Critics have argued that such attacks would have weakened al-Qaeda forces, helping prevent American casualties. Hagenbeck later explained that he decided against extensive air strikes for two reasons. First, there were limited fixed targets and most of these, such as caves and tunnels, were impossible to see from the air. At the same time, Hagenbeck wanted to preserve any intelligence information discovered in the underground complexes, something that would have been impossible if the cave entrances had been blasted and sealed by precision-guided munitions.

Long after the battle was over, U.S. Air Force officers continued to complain about Hagenbeck, claiming he made a mistake by leaving airpower out of the loop, at least initially. Several officers said they didn't even know about Anaconda until a day before the operation began.

On March 2, 2002, hundreds of Lodin's Afghan militiamen accompanied by Green Beret teams Shark, Texas 14 and Cobra 72, kicked off Anaconda and began their drive to push the enemy through the valley. Chinook helicopters started lifting the soldiers from the 10th Mountain Division into their blocking positions.

Then all hell broke loose. As Apache attack helicopters came into the valley to provide cover for the assault troops they were met with volleys of anti-aircraft fire and RPGs. Five of seven Apaches were put out of commission in the battle's opening days. A group of about 80 soldiers from the 10th Mountain Division landed in their blocking position only to find the enemy waiting for them. In addition, the Shark A-Team came under heavy fire with 12 of its Afghan soldiers wounded.

As Cobra 72 and its 70 Afghan troops started moving towards their objective, they also came under devastating and unrelenting fire. As rounds were landing around their trucks, the men scrambled from the vehicles, believing they had come under a mortar barrage from al-Qaeda forces hiding in the mountains around them.

Green Beret Chief Warrant Officer Stanley Harriman had already been wounded along with three of his special forces comrades, but they had been able to take cover in a nearby ditch. As the firing continued to rain down on the men, Harriman dashed back to his truck to retrieve a radio and was fatally hit.

It was later discovered that the attack that killed Harriman and two of his Afghan troops was not from al-Qaeda mortars but from an AC-130H gunship that had been providing fire support and reconnaissance for Cobra 72. An investigation conducted by SOCOM would reveal that the gunship was covering the advance of the Green Beret and Afghan convoy when it had received a call for fire support from another ground unit and flew to respond to that mission.

In the meantime, Harriman's truck and several other vehicles with Afghan troops separated from the main convoy and moved ahead to a pre-planned position. When the AC-130, which had already been plagued with a series of problems with its navigation equipment, headed back to hook up with the Green Berets, its crew spotted Harriman's truck and the other vehicles on the road below and immediately assumed they were al-Qaeda. The gunship opened up on the trucks with its cannon and machineguns.

The first rounds blasted into the front left side of Harriman's truck as the Green Beret sat in the front passenger seat using the radio. A second burst sent shrapnel through his legs and chest, while ripping off two of his fingers. More shrapnel entered through the back of his head.

Harriman's widow, Sheila, would later tell the CBS investigative TV show, *60 Minutes*, that her husband was on the radio listening in as the AC-130 gunship crew described the vehicles they were attacking on the ground. She said he tried in vain to radio a ground controller to abort the attack. "You're describing us," he frantically yelled into the radio before being hit. "You're describing us."

To make matters worse, Harriman's widow would also be told by his fellow Green Berets that their vehicles had been clearly marked as friendly forces.

By the time the AC-130 finished its attack run on the trucks, 34-year-old Harriman, who was assigned to the 3rd Special Forces Group at Fort Bragg, North Carolina, had become Anaconda's first American casualty. Three other Green Berets had been injured. Two Afghans were also killed and 14 wounded in the attack.

With the convoy shot up, one of his Green Beret advisors dead and his own men badly shaken by the gunfire, Lodin decided to withdraw his force from Anaconda.

Back in Bagram, U.S. commanders were surprised by the extent of enemy resistance they were encountering at some of the blocking positions and decided to send in a reconnaissance team from Task Force 11 to get a better idea of what was happening on the ground. The special operations team was to helicopter near a landing zone, dubbed Ginger after porn star Ginger Lynn, and set up an observation post on Takur Ghar, a 3,400-metre mountaintop. The peak was the perfect site for an OP with an unobstructed view of the southern approaches to the valley.

An MH-47E Chinook helicopter, code-named Razor 3, was to insert one team on Takur Ghar. A second team on Razor 4 would be inserted in another part of the valley. The 160th Special Operations Aviation Regiment, 2nd Battalion, would provide the helicopters while SEALs and Air Force combat controllers formed the reconnaissance teams.

At around 3 a.m. on March 4, Razor 3 approached a landing zone located on a small saddle atop Takur Ghar. The rear ramp door of the chopper was opened and the SEALs, including Petty Officer 1st Class Neil C. Roberts, got ready to step out into the frigid morning air. But as the MH-47 came in for its final approach, both the pilots and the men in back of the aircraft could see fresh tracks in the snow, indicating that al-Qaeda could be in the area.

Before they could abort the landing, the dark sky lit up with blasts from machinegun fire coming from every direction. An RPG tore into the Chinook's cargo bay but failed to explode. From 50 metres away a heavy machinegun riddled the Chinook's aluminum skin with bullets. "Get the fuck out of here!" one of the commandos yelled as the helicopter pilot gave full throttle to the engines in an attempt to get away from Takur Ghar.

As the Chinook was lifting off, the large aircraft was shuddering and shaking from the amount of damage it had sustained during its brief time over the landing zone. Al-Qaeda machinegun fire had severed the MH-47's hydraulic lines and electrical cables and the pilot struggled to keep the helicopter flying.

What happened next remains unclear. One version of the events recounted by military officers is that as the Chinook quickly lifted off from Takur Ghar, Roberts, who was firing his machinegun from the back ramp now covered with leaking hydraulic fluid oil, slipped out of the helicopter and fell onto the ground. A crew chief on the aircraft tried to grab Roberts but also tumbled off the Chinook. That man spent several terrifying moments dangling behind the helicopter, suspended in the air by his safety line, before being dragged back in by another crew member.

Some SEALs, however, put forward a different version of the event. They say that it was Roberts trying to pull the crew member in when an RPG hit the nose of the Chinook but failed to explode. Roberts was knocked off balance and tumbled out, falling about three metres to the ground.

An official Pentagon report on the battle simply states that Roberts and the crew member at the rear were trying to steady each other when they both slipped on the oil-soaked ramp and fell out of the Chinook.

With the electrical systems shutting down and hydraulic fluid draining from the aircraft, the controls on the MH-47 started seizing up. It would take all the pilot's flying skills to keep the limping chopper airborne and put it down for a controlled crash-landing some seven kilometres north of Takur Ghar.

The SEALs on board quickly did a head count, confirming what they already knew - Petty Officer Roberts was missing.

Forty minutes later, the Chinook crew and SEALs were located by their comrades in Razor 4 which had already completed its mission and dropped off its special operations team at another location. With Razor 3 disabled, Razor 4 now had the SEALs on board and two full Chinook crews, too much weight for it to operate in the thin mountain air and climb back up to the location where Roberts was last seen. Instead, the decision was made to fly back to a staging area at Gardez, drop off the extra Chinook crew and then head out to rescue the SEAL.

As Razor 4 touched down to drop off the extra helicopter crew, a Spectre gunship was circling over Takur Ghar trying to determine the situation on the ground. Scanning the terrain with its surveillance sensors, the crew thought they could see a lone individual, later believed to be Roberts, surrounded by four to six people.

Also high above the site was a Predator UAV, transmitting live, but fuzzy, images back to Hagenbeck's command centre in Bagram. Officers there watched with grim faces as what appeared to be a group of al-Qaeda approached Roberts and dragged him away.

At 5 a.m., Razor 4 was coming in for its approach on Takur Ghar, its pilots prepared for the worst. They didn't have to wait long. When the helicopter was 10 metres from the ground, they could see the muzzle flashes from al-Qaeda machineguns. The rounds tore into the chopper and smashed into the front windshield. Inside, a crew chief could hear the sound of pings and pops as the bullets impacted against the Chinook and cut through its skin.

As soon as the MH-47 touched down, the special operations troops charged off its back ramp into the snow to begin their search for Roberts. The Chinook, although heavily damaged, was able to lift off.

Back at Bagram airbase, Hagenbeck watched intently as a Predator UAV transmitted live images of the attack. He couldn't see the al-Qaeda hidden in trenches or caves, but he watched the special operations forces, in an act of incredible courage, rushing straight forward under intense fire towards the enemy.

The most prominent features on the mountaintop and the ones the men headed for were a large rock and tree. As the team approached the tree, Air Force combat controller Technical Sergeant John Chapman could see two al-Qaeda in a bunker. He opened up on the men with his M4 carbine, killing both.

Al-Qaeda in another enemy bunker about 20 metres away immediately opened fire on Chapman and the SEALs with him, killing the 27-year-old combat controller. The SEALs peppered the bunker with gunfire and tossed grenades into the fortification.

Clearly outnumbered, with two SEALs wounded and Chapman dead, the special operations troops decided to break contact with the enemy. They killed two more al-Qaeda and, with one of the wounded SEALs taking point, started to head down the mountain.

As they withdrew, they radioed for fire support from a gunship, call sign Grim 32. The AC-130 was more than happy to oblige, raking the mountaintop with a 105 mm howitzer and 40 mm cannon fire.

As the SEALs made their way down to safety, unknown to them, a rescue force of 23 U.S. Army Rangers and Air Force special operations troops on board two Chinooks was already enroute to Takur Ghar. One of the Rangers' Chinooks, Razor 1, came into the landing zone around 6:30 a.m. only to be greeted by a volley of RPG-7 and machinegun fire from the al-Qaeda force still on the mountain. A rocket-propelled grenade tore into the Chinook's right engine, causing the chopper to drop like a rock in the last three metres of its descent. Another RPG whizzed through a window on the aircraft's right side, starting a fire.

On board, U.S. Air Force combat controller Sergeant Gabe Brown could see the rounds coming into MH-47, shredding the chopper's insulation and turning it into a kind of confetti which filled the cabin. "Here we go," Brown said to himself.

Just a couple of hours before, the sergeant had been sleeping when someone shook him awake. "Start spooling up," he was told. "A helo is down."

Now he was in the middle of an al-Qaeda ambush.

The Chinook had put down on a flat area along the ridge on Takur Ghar. On the other side was a cliff face dropping off about 300 metres. Communications problems meant that Razor 1's pilots didn't know in advance that al-Qaeda were swarming all over the landing zone. Bullets tore into the cockpit, slamming into the legs of one of Razor 1's pilots. Other men were gunned down as they exited the MH-47.

Moving quickly from the disabled Chinook, Air Force Staff Sergeant Kevin Vance saw the carnage al-Qaeda forces had inflicted. The helicopter's door gunner was lying on the aircraft's back ramp, an AK-47 bullet in his head. A second person was at the end of the ramp face down in the snow. He had been shot in the chest. A third dead man was sprawled on the ramp lying on his back. Another Ranger had been hit while still inside the aircraft and killed instantly.

Although he didn't realize it at the time, 25-year-old Vance had also been hit by shrapnel in his shoulder.

The men took up positions in the snow and began firing at the al-Qaeda scattered around the landing zone. Most of the Rangers' efforts, at first, were concentrated on the bunkers some 50 metres away. Back inside the Chinook, Senior Airman Jason Cunningham, a pararescue specialist, along with another medic, pulled off the remaining insulation from the aircraft to wrap the casualties in so they would stay warm.

As the al-Qaeda continued to fire, Sergeant Brown radioed to any aircraft in the area for help. He was able to raise two F-15 fighter jets which soon came roaring into the valley, raking enemy positions with their 20 mm gatling guns. Brown could see the snow flying off the ground near one of the al-Qaeda bunkers, indicating that the F-15 attack was on the mark. But even after the fighters made several passes, emptying all their ammunition onto the enemy positions, Brown and Sergeant Vance could see that some of the al-Qaeda were still jumping up and shooting.

Several tossed grenades at the Americans but these fell three metres short and, buried in the snow, exploded harmlessly. Brown proceeded to call in more aircraft and at times their bombs dropped as close as 50 metres from the Rangers.

In other observation posts on adjoining mountaintops, Australian SAS and American special operations forces from Task Force K-BAR could see al-Qaeda below Takur Ghar climbing up to reinforce their comrades. Throughout the day, the commandos called in air strikes on those enemy forces, preventing the Rangers from being overrun.

The other half of the Ranger rescue force, who were on board a Chinook code-named Razor 2, had been ordered to move to a safe area and await further instructions after it was determined that Razor 1 had gone down. Razor 2's pilots eventually decided to return to the Gardez staging area where there, after a heated argument on the ground and threats from one of the Rangers on board, the crew finally agreed to take the rescue team back up the mountain.

The Chinook inserted its 10 Rangers and one SEAL at about 8:30 a.m., but the pilot dropped them off 700 metres below their comrades who were still fighting on the top of Takur Ghar. Since the men would have to climb up a steep mountain slope covered in places by waist-high snow, the first order of business for the Rangers from Razor 2 was to dump all unnecessary gear, taking only ammunition and weapons with them.

From his position on top of Takur Ghar, Sergeant Vance could see the Razor 2 rescue force climbing towards him as al-Qaeda started lobbing mortar bombs down on them. It would take two hours but the exhausted Rangers from Razor 2 were eventually able to reach the top of the mountain and link up with their comrades. There, for the next 12 hours, the two groups of Rangers and their Air Force special operations comrades fought off the al-Qaeda attacks.

With Sergeant Vance's direction, Sergeant Brown continued to call in air strikes on the fortified enemy positions. In one instance, a massive 1,000-pound bomb landed just 150 metres away from the group and Vance watched as flaming debris rocketed over the Rangers' heads and down into the valley where it ignited trees. The blast from a 1,000-pounder in such close proximity was too risky and this was the last time such a heavy bomb was used.

Despite the bombing, the al-Qaeda still controlled the top of the mountain from their bunkers, a mere 50 metres away from the Rangers. With the arrival of the men from Razor 2, the Rangers prepared to assault the enemy stronghold. Brown called in a last air strike on the al-Qaeda positions and then seven Rangers stormed the bunkers, moving in knee-deep snow as they fired their M4s and tossed grenades. The Rangers were able to overrun the fortifications and kill the al-Qaeda inside. In one bunker, they found Chapman's body. Roberts was discovered about five metres away.

Just when it looked as though the battle had subsided, two more al-Qaeda appeared about 400 metres behind the Rangers and opened up on a group of wounded men who were near the downed chopper. Cunningham, the pararescue specialist, was dragging an injured man to cover when he was hit in the pelvis and abdomen by a stream of bullets. The 26-year-old airman patched up his own wounds and continued to treat other Rangers, but several hours later he slumped into the snow, dead from his own injuries.

Shortly after 8 p.m., the shooting dwindled and three helicopters managed to make it onto Takur Ghar. The Rangers scrambled on board the aircraft while a fourth chopper headed to a site lower down the mountain where it picked up the SEAL team that had originally tried to rescue Roberts.

Anaconda would continue for another week as al-Qaeda reinforcements moved into the area. Swarms of A-10 tank buster aircraft would scour the mountains, using their gatling guns to cut down fleeing al-Qaeda. A single crew was credited with at least 100 kills. One al-Qaeda, either brave or foolish, rose out of his foxhole with a RPG-7 to confront three Cobra attack helicopters. He was shredded to pieces by the helicopters' gatling guns before he could fire his rocket. The Green Berets of Cobra 72 went back into action, calling in air strikes against al-Qaeda bunkers. New Afghan troops, these from the more reliable Northern Alliance, arrived with T-55 tanks to join their special operations comrades in the Shah-e-Kot area.

America's coalition allies also got in on the action. Early on in the fighting, two Canadian Army sniper teams, outfitted with .50 calibre McMillan rifles, worked with U.S. SOF to hunt down al-Qaeda who tried to ambush American paratroopers as they moved into their blocking positions. The Canadians were credited with more than 20 confirmed kills during Anaconda. In one unconfirmed kill, a Canadian sniper shot an al-Qaeda 2,400 metres away as the man was driving a resupply truck into the Shah-e-Kot Valley.

But as the days wore on, contact with the enemy lessened. On March 13, soldiers from the 3rd Battalion, Princess Patricia's Canadian Light Infantry were brought in to conduct a sweep along a mountain ridge called the Whale's Back. They found large numbers of caves and built-up fighting positions but the only enemy they discovered were two al-Qaeda hold-outs in a bunker. A team from the 10th Mountain Division was called in and destroyed the fortification with two anti-tank rockets. The enemy fighters didn't know what hit them; the blast from one of the AT rounds decapitated one of the al-Qaeda as he was sitting down to eat a bowl of rice.

Anaconda officially ended on March 19, leaving commanders to count up their losses and try to figure out what went right and what went wrong. Dozens of American soldiers who fought at Ginger and on the top of Takur Ghar, as well as at other locations, had been wounded. Eight U.S. service men – seven in various battles at Takur Ghar, and Harriman with Cobra 72 – died during Anaconda.

John Chapman, the combat controller killed after heading back for Roberts, was posthumously awarded the Air Force Cross, his service's highest award for heroism. In the citation for the award, Chapman was credited with killing two of the enemy as he charged the al-Qaeda machinegun emplacement.

"His conduct culminated in his selfless decision to place his own life in jeopardy to save fellow warriors, and thus by his sacrifice, we the living are called upon to adhere daily to those values this country cherishes," Air Force Secretary James Roche said of the sergeant's actions.

Senior Airman Jason Cunningham also received a posthumous Air Force Cross while others involved in the firefight at Takur Ghar were awarded a variety of medals, including Silver Stars, Distinguished Flying Crosses and Bronze Stars.

The last moments of 32-year-old Petty Officer Neil Roberts remain a mystery, though in each version of the events leading to his death the SEAL was said to have fought bravely as he was overcome by far superior numbers of al-Qaeda.

Initially, some officers believed that Roberts was killed almost immediately by al-Qaeda who descended upon him when he fell out of the helicopter. But some of his fellow SEALs believe Roberts, armed with his M249 Squad Automatic Weapon, died after a fierce gun battle that lasted anywhere from 90 minutes to two hours. They say after he fell from the Chinook, he low-crawled to a defensive position and began opening fire on advancing al-Qaeda. Roberts died as he launched an attack on an enemy machinegun nest, according to the SEALs.

Forensic evidence gathered at the scene by SOCOM determined that Roberts survived the fall and then turned on an infrared strobe light he was carrying so rescue forces could find him. The videotape from the Predator UAV also provided some clues. "The image was fuzzy, but we believe it showed three al-Qaeda had captured Roberts and were taking him away around to the south side of Ginger and disappearing into a tree line," Hagenbeck told *The Washington Post*. "That was 15 to 20 minutes before the first rescue team arrived."

SOCOM investigators concluded from their evidence that Roberts used his SAW to try to fight off the enemy but the gun may have jammed. The weapon, covered in blood, was found near his body with unfired rounds in the chamber. Investigators also concluded that the SEAL had been shot at close range. He would be posthumously awarded a Bronze Star for his "zealous initiative, courageous actions, and exceptional dedication to duty."

Shortly after his death, Roberts' family released to the news media a letter he had left to be opened in the event he was killed in Afghanistan. They wanted the public to know that Roberts was a loving father to his 18-month-old son and that being part of the special operations community was one of the highlights of his life. "I loved being a SEAL," Roberts wrote. "If I died doing something for the teams, then I died doing what made me happy. Very few people have the luxury of that."

For the Pentagon, Anaconda was considered a victory. It estimated that anywhere from several hundred to 700 al-Qaeda were killed during the operation. Most of the dead were from Uzbekistan but there were some Chechens and Chinese Islamic fighters as well.

But Afghan militia commanders and locals in the region believe that the majority of the al-Qaeda force, numbering anywhere from 500 to 1,000, slipped over the Pakistani border when they were tipped off about the American offensive. The al-Qaeda that coalition troops did fight were mainly a rear-guard willing to be martyred, those people say.

CENTCOM commander General Tommy Franks and Major General Hagenbeck dismissed the suggestion that the main al-Qaeda force escaped before Anaconda began. At best, only small numbers of the enemy were able to get out of the valley alive, they maintain.

Franks also defended Anaconda against criticism that U.S. intelligence was flawed and had seriously underestimated the number of al-Qaeda in the area. He said the intelligence was the best available at the time, although he acknowledged it was far from perfect. Franks said of Anaconda: "It is the stuff of which heroes are made."

For his part, K-BAR's commander Captain Harward saw the operation as a success and one that forced al-Qaeda and the Taliban to go on the run, turning them from a cohesive force into scattered and uncoordinated units.

Again, the study by U.S. Army War College research professor Stephen Biddle, as well as an official Pentagon investigation, provided valuable insight into possible reasons why the number of enemy may have been underestimated by American commanders. The Pentagon report noted that some of the al-Qaeda positions, such as a command post among the boulders, were almost impossible to see from the air. The command post, a green tent set between a crevice of rocks, looked like a shadow on the mountain to anyone looking at it from above. As well, blowing snow meant that the footprints of al-Qaeda soldiers would be quickly covered.

Likewise, Biddle, through interviews with the American combatants, found that the al-Qaeda bunkers were virtually impossible to see from the air. One dug-in command post would be pounded by five JDAMs yet the enemy inside survived and were only killed when they were overrun by U.S. troops, he discovered.

At the Ginger landing zone, the main al-Qaeda bunker was not even visible at close range from the ground, according to Biddle's report. Its overhanging rock provided cover and concealment. Even if special operations troops were able to observe resupply movements or al-Qaeda patrols, the enemy was dressed similarly to local Afghan herdsmen who tended goats in the valley and it would have been difficult to distinguish the two, Biddle concluded.

Above all, the terrorist force that met the special operations and conventional troops was not a ragtag army of Afghan tribesman whose loyalty could be bought by the highest bidder. These were well-trained Islamic fighters, motivated by their desire to kill Americans. This was the battle they had hoped for, Biddle noted. "Al-Qaeda defenders not only stood their ground against overwhelming American firepower, they actually reinforced their positions in the midst of the battle," he wrote. "Their fighters were willing to advance into the teeth of a fierce bombardment to enter the Shah-e-Kot Valley from safer positions elsewhere and seek battle with our forces."

In the weeks following Anaconda, K-BAR's mission would wind down, but not before the SEALs suffered another fatality. On the morning of March 28, SEAL Chief Petty Officer Matthew J. Bourgeois was conducting a breaching exercise at Tarnak Farm, six kilometres south of Kandahar Airport, when he stepped on a landmine. The farm, once an al-Qaeda training camp, had been taken over by coalition forces for their own training activities. The blast killed the 35 year old and wounded another SEAL. Bourgeois, a 14-year naval special warfare veteran who left behind a wife and seven-month-old son, was a month away from coming home when he died.

Other K-BAR personnel had close calls with death. In one incident, an MC-130E aircraft plowed into the side of a mountain right after it finished an aerial refueling in poor weather. Incredibly, the snow was so deep, easing the impact of the crash, that the crew walked away with minor injuries.

Despite such setbacks, during its time in Afghanistan, K-BAR inflicted much more damage on the enemy than the losses it sustained. In addition, its operators had managed to gain valuable intelligence on al-Qaeda's network. During its raids on caves and tunnels, K-BAR operators found that their terrorist enemy was well-equipped. Among the equipment captured were laptop computers, satellite phones and U.S. military-issue PRC-117 radios.

Afghanistan also validated the SEALs' intense training regime. The teams' emphasis on top physical conditioning ensured they were able to operate high in mountains and valleys without suffering from altitude sickness. The climbing and rappelling skills they learned for conducting assaults on boats and offshore oil rigs were put to use in establishing observation posts on high-altitude peaks.

Back in the U.S., Commander Kerry Metz, K-BAR's director for operations, briefed American politicians on the task force's accomplishments.

He described how special reconnaissance teams were operating in the mountains of Afghanistan above 3,500 metres for extended periods without re-supply. Every trick of the trade, every piece of equipment and weapon was put to use to inflict the maximum amount of damage possible on the enemy.

"We challenged our operators to conduct missions in some of the most hostile environments ever operated in," Metz said. The K-BAR operators, he concluded, were among the finest in the world.

In the end, K-BAR earned an impressive list of accomplishments in Afghanistan. The task force had completed 42 special reconnaissance missions and 23 direct action or sensitive site exploitations. It captured 107 enemy soldiers, called in 147 air strikes for close air support or to strike at targets of opportunity, and destroyed more than 225,000 kilograms of enemy explosives and weapons. Confirmed enemy dead were at least 115.

PREVIOUS PAGE: *The al-Qaeda complex at Zhawar was massive and well-constructed. In this photo, SEALs pass through one of the elaborate cave entrances during their January 2002 mission to the underground labyrinth.* (COURTESY USN)

ABOVE: *U.S. paratroopers search for enemy forces on ridges above them during Operation Anaconda.* (COURTESY U.S. ARMY)

LEFT: *U.S. SOF in civilian clothing move along the mountain slopes during Operation Anaconda.* (COURTESY DOD)

TOP: SEALs, attached to the USS Shreveport, search a ship for fleeing al-Qaeda or Taliban in the Arabian Sea in December 2001. (COURTESY USN

ABOVE LEFT: SEAL Neil Roberts was killed during fighting at Takur Ghar in March 2002. (COURTESY USN)

ABOVE RIGHT: SEAL Matthew Bourgeois was killed after stepping on a landmine during a training exercise near Kandahar in March 2002. (COURTESY USN)

A U.S. Army Green Beret and an Afghan militiaman working together in the early days of the Afghan war. (COURTESY SOCOM)

U.S. SOF with Afghan civilians during Operation Enduring Freedom. (COURTESY SOCOM)

A room with a view: a high-altitude JTF2 observation post in the Afghan mountains.
(COURTESY JTF2)

JTF2 soldiers from Task Force Canada on a night mission in Afghanistan. The operator on the right has a C9 machinegun with its stock removed.
(COURTESY JTF2)

TOP: *A SEAL gunner on a DPV (desert patrol vehicle) near Kandahar International Airport, February 2002. (COURTESY USN)*

BOTTOM: *U.S. SOF – likely Green Berets and CIA operatives – ride into battle on horseback during the early days of the Afghanistan war. (COURTESY DOD)*

ABOVE: A U.S. SOF operator armed with an M4 carbine during Operation Enduring Freedom. (COURTESY SOCOM)

MAIN: These two Australian SAS wear the unit's distinctive desert camouflage uniforms in Afghanistan. (COURTESY ADF)

*An Australian SAS trooper on a 6X6 all-terrain vehicle meets up with his comrades who are in one of the unit's long-range patrol vehicles. (COURTESY ADF) / **BELOW:** SEALs watch as an enemy munitions cache is destroyed. (COURTESY USN)*

ABOVE: *An Australian SAS patrol finds out firsthand just how miserable Afghan winters can be.*
(*COURTESY ADF*)

LEFT: *An Australian SAS fires his M4 at a range near Bagram.*
(*COURTESY ADF*)

TOP: *A SEAL on patrol somewhere in Afghanistan. He is armed with an M4, which is outfitted with the rail system to attach various options such as sights and laser-targeting devices.* (COURTESY USN)

ABOVE: *SEALs examine munitions captured during one of their raids in Afghanistan.* (COURTESY USN)

U.S. special operations personnel get ready to climb aboard a helicopter for a mission in Iraq.
(COURTESY DOD)

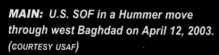

MAIN: *U.S. SOF in a Hummer move through west Baghdad on April 12, 2003.* (COURTESY USAF)

BOTTOM LEFT: *Australian 4RAR (Commando) soldiers move cautiously among buildings at an undisclosed location in Iraq. The 4RAR was a quick reaction force for the Australian SAS.* (COURTESY ADF)

BOTTOM MIDDLE: *A SEAL provides close protection for senior U.S. officers in the Philippines. SEALs and other U.S. SOF have been training Philippine military forces who are fighting guerrillas linked to al-Qaeda.* (COURTESY USN)

BOTTOM RIGHT: *Soldiers from the 4th Battalion, Royal Australian Regiment (Commando), accompanied Australian SAS during the raid on Al Asad airbase. Much of Iraq's air force was discovered hidden at the base.* (COURTESY ADF)

An Australian SAS trooper somewhere in western Iraq. Note the suppressor on the end of his rifle. (COURTESY ADF)

TOP: *A U.S. Air Force combat controller provides security while his teammates inspect a building near Baghdad on April 29, 2003.* (COURTESY USAF)

ABOVE: *An aerial gunner on an MH-53M Pave Low IV helicopter mans his .50 calibre machinegun during a combat mission in support of Operation Iraqi Freedom.* (COURTESY USAF)

MAIN: Polish GROM take up defensive positions during boarding operations in the port of Umm Qasr in March 2003. (COURTESY USN)

RIGHT: A SEAL covers his team members as they prepare to board a ship during a training mission in the Persian Gulf. (COURTESY USN)

ABOVE: *Australian SAS troopers on patrol in northwestern Iraq. They are equipped with M4s outfitted with suppressors. (COURTESY ADF)*

RIGHT: *Spanish SOF fast-rope onto the North Korean freighter So San during a December 2002 mission. The team found 15 Scud missiles hidden on the ship. (COURTESY SPANISH DEFENSE MINISTRY)*

OPPOSITE PAGE: *Australian SAS patrol in the Afghan desert. (COURTESY ADF)*

Diggers at War

4 To the Afghans, the strange vehicle rumbling through their village must have looked like the ultimate war machine.

In a turret above the vehicle's driver, a muscular man with sunglasses stood behind a .50 calibre machinegun. Beside the driver, another man, his face partially covered with a scarf, wielded a 7.62 mm calibre MAG 58. Ammunition containers were packed into the back of the six-wheeled truck and camouflage netting was draped on its hood. Other men inside the vehicle carried M4 carbines. Sometimes the trucks were accompanied by heavily armed motorcycle riders, who tore ahead across the desert to search out the enemy.

The Afghans' reaction to these patrols by the Australian Special Air Service was mixed. The kids loved to see the commandos and would often run alongside the vehicles trying to get their attention. Some of the adults waved at the SAS while others would glare or spit on the ground as they passed.

As the Australian patrols rolled through villages and towns, the SAS men were struck by the large number of civilians carrying weapons. It wasn't unusual to see a utility vehicle full of RPG-7 projectiles drive by one of the patrols. Teenagers wandered through the streets armed with AK-47s. Figuring out which of the population was friend or enemy was just one of the many challenges of operating in Afghanistan.

The main purpose of the SAS patrols was to get into the country's remote areas in the south, especially those not previously explored by U.S. forces. Some missions involved conducting clandestine strategic reconnaissance aimed at gathering intelligence about al-Qaeda and Taliban movements. Such operations were quickly followed by air strikes or lightning-quick raids to uncover weapons caches or search caves. Other patrols were overt, a heavily armed display designed to announce the Australian presence in the area.

Whatever the mission, the message was obvious: the Diggers, the Australian nickname for their fighting men, were not to be messed with.

The Australian SAS had been assigned to Captain Robert Harward's Task Force K-BAR. The officer was confident enough in the Australians' abilities to give them wide latitude in conducting their patrols. And it was a credit to the SAS Regiment that their participation in the Afghanistan war was considered by key U.S. decision-makers soon after the September 11 attacks.

In fact, it was three weeks after the attacks that the idea of having the Australian SAS participate in the planned U.S. offensive against al-Qaeda was first raised at a meeting in Washington. Bush administration officials were still developing their strategy on how to launch a decisive blow against the terrorist network and, particularly, how to get U.S. SOF on the ground in Afghanistan. Defense Secretary Donald Rumsfeld and Secretary of State Colin Powell were reviewing the progress so far; planning was indeed slow going.

Condoleezza Rice, President Bush's assistant for National Security Affairs and one of his closest advisors, told the men that U.S. allies were eager to contribute. Canada had offered naval forces, at least for starters. The French were talking about fighter jets as well as ships. Even Germany, a country known for its reluctance to deploy troops beyond its own borders because of its war-time past, wanted a piece of the action.

But, in particular, Rice singled out Australia's special forces. Two Australian SOF officers were already working at General Tommy Franks' CENTCOM headquarters in Tampa, Florida, she noted. What about giving Australian SOF a role?

Australian Prime Minister John Howard had given a speech just weeks before stating that not only was his country firmly behind America, it was ready to contribute military forces. Howard's recommendation to U.S. officials was that any response should be "targeted yet lethal."

That phrase was enough to start speculation in Australia that the country's SAS would play a role in any coalition war effort.

For his part, Rumsfeld was reluctant, at first, to consider allied participation. To be sure, having more nations on board the U.S. coalition was a good diplomatic and public relations strategy. But Rumsfeld was worried that having to consult with other countries could get in the way of the expeditious military onslaught he was about to release on the Taliban and al-Qaeda. First and foremost, any foreign contributions would have to make military sense.

Rumsfeld needn't worry about that problem with the Australian SAS.

Established in 1957 and modeled on the British SAS, the 600-to 700-strong regiment had earned a solid reputation fighting guerrillas in Borneo and Vietnam. It had a respectable combat record in Southeast Asia where it had worked alongside American troops. In its six years conducting reconnaissance and ambushes in Vietnam's jungles, the SAS had confirmed kills of 492 enemy soldiers and another 106 possible kills. The cost to the regiment was one of its men killed in action and 28 seriously wounded.

More recently, the regiment's overseas deployments had been sporadic, with its operators serving in Somalia and East Timor and on various short-term missions where it appeared Australian citizens would have to be evacuated from war-torn countries.

The initial stages of this new American war on terrorism would be fought in Afghanistan, a country known for its deserts. The Americans had SOF units that could conduct long-range reconnaissance patrols in such terrain. But the Australians were acknowledged as one of the leaders in this area, having years of experience patrolling the remote regions of their own country.

Over the years, the unit – based at Campbell Barracks, Swanbourne, near Perth – had also built up strong links with the U.S. special operations community. Early in 1998, when American commandos were deploying to Kuwait to take part in a possible attack on Iraq, the Australian SAS were among the first foreign SOF units that U.S. commanders requested.

In that operation, dubbed Desert Thunder, the Americans were flexing their military muscle along Iraq's border after Saddam Hussein refused to allow United Nations inspection teams into the country. Air strikes were being planned against Iraq and the Pentagon wanted the Australian SAS in place for the rescue of any downed British and American pilots behind enemy lines. That February, 200 to 250 SAS troopers were deployed to Kuwait to take part in the mission, but weeks later a political compromise was worked out by the UN and the situation was defused.

The SAS, however, stayed on for several months and, later, under the command of Lieutenant Colonel Rowan Tink, who would also be picked for the Afghanistan mission, the troopers conducted desert patrols and helicopter insertions using U.S. MH-47s.

The SAS learned some valuable lessons while taking part in the U.S.-led Desert

Thunder coalition, a code-name later changed to Desert Spring when it became apparent that air strikes had been put on hold. During the operation, the Australians were able to improve their liaison with their American counterparts, sending two special operations officers to work at CENTCOM in Florida. Desert Spring also made it clear that some equipment needed to be upgraded, so the regiment purchased long-range patrol radios as well as adopting the U.S.-made M4 carbine.

All would be put to use in the Afghanistan mission. Less than a month after the September 11 attacks, the Australian government officially announced the deployment of its forces to support the U.S. war effort. Almost 2,000 personnel on ships and aircraft would be assigned to the theatre of operations. But Australia's most important commitment – not in numbers but certainly in the effect they would have on the ground – would be 150 of the country's SAS troopers.

In preparation for the mission, the SAS unit was issued with the Australian Army's new disruptive pattern camouflage for use in desert conditions. The uniforms were quite distinctive, being a sand-colored tan with varying shades of brown dots scattered throughout. The troops were also issued with cold-weather clothing to deal with the coming Afghanistan winter, as well as Javelin anti-armor missiles for extra firepower.

Giant Russian-made Antonov cargo planes were rented to move the regiment's customized patrol vehicles to Southwest Asia. The SAS took both its six-wheel drive Long Range Patrol Vehicles (LRPVs) and four-wheel drive Land Rover 110 Series Regional Surveillance Vehicles (RSVs). Although there were some in the regiment who weren't enamored of the trucks, believing they drew undue attention to SAS operations, the "Mad Max"-style war wagons would come in handy in Afghanistan's vast deserts and desolate terrain.

Firepower on the LRPV, equipped with a .50 calibre machinegun as well as a 7.62 mm MAG 58 general purpose machinegun, could be augmented by an automatic grenade launcher or a Javelin. The RSV had the general purpose machinegun as well as a Minimi machinegun. For shorter patrols, the regiment also packed its six-wheel all-terrain vehicles and its moto-cross bikes.

The Australians started arriving in Afghanistan on November 27 with their first location at Forward Operations Base Rhino, some 100 kilometres southwest of Kandahar. A much larger group from 1 Squadron moved from its staging area in Kuwait to Rhino on December 2.

The Australians shared the base with the U.S. Marine Corps and they quickly became the eyes and ears for the Americans, establishing whether al-Qaeda or Taliban were operating in the area.

Five- and six-man SAS patrols made their way across the Afghan desert and mountains. The country lived up to its reputation as being one of the most inhospitable places on earth. Dust storms were a constant problem, covering patrol vehicles and SAS operators with a thin coat of sand. At night, the temperature dropped to minus 15 Celsius, and was even colder when the wind chill was factored in. At times, drizzling rain or snow flurries turned dirt roads into mud slicks. Hard rations were standard on the patrols and water was carefully conserved, being the most scarce commodity on a mission.

During the early days of the deployment, SAS patrols provided a constant flow of information on al-Qaeda and Taliban movements along the key roads around Kandahar and in the Helmand Valley near the Iranian border.

On December 21, the SAS relocated from Rhino to the airport just outside Kandahar and soon began vehicle patrols from their new base. Around Christmas, one of the SAS reconnaissance teams discovered a hidden Taliban fuel and logistics dump. It was put under observation for a day or so before the SAS called in an air strike to destroy it.

Just days later, another patrol was sent on a reconnaissance mission to scope out a network of tunnels and caves thought to house an al-Qaeda training base. The SAS, along with U.S. special operators, moved into the complex on December 28. Although they discovered large amounts of weapons, explosives and ammunition, they found that the enemy had already cleared out. The complex was so large it would take several days to properly search the site.

Another sensitive site exploitation mission involved a compound near Lashkar Gah, one of the residences of Taliban leader Mullah Omar. Initially, the location was put under surveillance. Once the SAS moved in, they checked the site for intelligence information as well as chemical and biological agents.

The missions in December set the tone for what was to follow. Enemy ammunition dumps and arms caches were being discovered, but the Taliban and al-Qaeda had left long ago.

"So these things, I won't say keep turning up with monotonous regularity, but they're out there and it's a matter of essentially finding them," Brigadier Gary Bornholt, the Australian contingent commander, explained at one briefing in January 2002.

Most of the missions the SAS were conducting in southern Afghanistan were covert and leading to a significant amount of quality intelligence, according to Bornholt. The SAS patrols would set up at a "hide" location and then operate from there for several days before moving on. The idea was to keep the enemy off balance, not knowing where the SAS would turn up next. After Australia's participation in the war was over, Brigadier Duncan Lewis, head of the country's special operations command, would reveal that one SAS patrol spent almost three months constantly deployed in the field, moving from location to location.

But, in fact, it seemed that al-Qaeda was the one keeping the coalition forces guessing about where its operatives were. A 10-day sweep by the SAS of an area 60 kilometres north of the town of Ghazni came up dry. Other operations involved more sensitive site exploitation work, checking al-Qaeda caves and bunkers, all of which were abandoned.

One of the main dangers the SAS faced was from landmines and booby-traps. Not only had the Taliban and al-Qaeda rigged some of their compounds with these nasty surprises, but the country was covered with several million landmines planted by the Soviets during their occupation of Afghanistan.

The landmine threat finally caught up with the SAS shortly after noon on January 17, 2002, when one of its soldiers had two of his toes blown off after he stepped on an anti-personnel mine. The SAS trooper had been on a sweep north of Kandahar when the incident occurred. SAS medics treated the man and a U.S. Black Hawk medevac chopper was airborne within 10 minutes of receiving a request for help. The trooper's patrol discovered a large arms cache in the area where he was hurt and as much as 15 tonnes of ammunition was destroyed.

By late January, the SAS had conducted 12 operations ranging in duration from a few days to several weeks. Although the commandos discovered weapons, ammunition and intelligence documents, they still hadn't had contact with the enemy. That, according to Brigadier Bornholt, was not necessarily a sign that the missions were not going right.

"In a lot of cases, if our special forces soldiers actually have to resort to engaging the adversary with weapons, then to a large degree they've probably failed the mission," he explained. "Because they won't have been able to observe and gain this information that we need from a clandestine perspective which is where we get our greater strength."

But some of the SAS men in Afghanistan were clearly frustrated by the lack of actual fighting. "It's not the war they promised," an unidentified trooper at the Kandahar base told journalist Craig Nelson of the *Sydney Morning Herald*. Days after Nelson filed his report, he was banned from the base, supposedly for revealing the existence of the Australian SAS at the installation, a fact that had already been well reported in the media.

SAS soldiers summed up the prevailing attitude in Afghanistan at the time by taping a cartoon on the door of one of their tents at Kandahar. One picture showed a maze of empty tunnels and bunkers, while the second showed Osama bin Laden leisurely pushing a shopping cart through a store in Perth, the SAS home base.

Like their American SOF counterparts, the Australians also had to deal with the inevitable questions from the press about the hunt for bin Laden. Where was the al-Qaeda leader and why hadn't he been caught?

"What we've said before is this is not us going after Ned Kelly," explained Bornholt, referring to the Australian folk hero and bank robber of the late 1800s. "It's about us going after outlaws. And rather than focus on one individual, you know the focus of our operation is to support the coalition and gain information so that they can take further action. I mean if this guy turns up then he's in a bit of trouble."

On February 16, 2002, the unit suffered its first fatality when one of its patrols drove over an anti-tank mine during a mission in the Helmand Valley. The massive explosion destroyed the patrol's six-wheel LRPV and blew off the leg of its leader, Sergeant Andrew Russell. Four other SAS troopers emerged from the blast dazed but uninjured. The survivors immediately began administering first aid to 33-year-old Russell and radioed for a helicopter to evacuate their casualty.

It took about 20 minutes from the time of that radio transmission until a U.S. helicopter lifted off to begin its flight to reach the patrol. When it became clear that the slow-moving chopper might not reach Russell in time, a three-man U.S. Air Force pararescue team boarded a fixed-wing aircraft and was able to arrive at the SAS location in about an hour. Ignoring the risk to their own lives, the three parachuted into the minefield. By the time the helicopter reached Russell, the U.S. Air Force pararescue crew had already begun to administer intravenous fluids to try to stabilize the sergeant's condition. Despite their best efforts, however, the veteran soldier died en route to the U.S. hospital in Kandahar.

Russell, who had joined the army when he was 20, left a wife and a two-week-old daughter whom he had never met. His life in the SAS was typical of many of the operators in the regiment. Besides serving in Kuwait and East Timor, many weeks were spent on grueling training exercises. It wasn't uncommon for Russell to be at home for fewer than six months of the year.

Russell's death was the first Australian war fatality since the Vietnam conflict. It sparked debate in the country's Parliament as to whether the SAS should rely on U.S. medical evacuation helicopters in the future, while also raising questions about whether the families of fallen special operations soldiers would be given adequate financial support.

Several days after the accident, Russell's body was put on a C-130 transport plane at Kandahar for return to Perth. It was a solemn ceremony as his fellow SAS operators watched his coffin being loaded onto the U.S. Air Force Hercules while a bagpiper from the Princess Patricia's Canadian Light Infantry played a lament.

But those at the base would have little time to mourn their fallen comrade. On February 20, special operations troops were amassing at the airbase at Bagram. Days later, teams from several countries began infiltrating into the Shah-e-Kot Valley for what was supposed to be a knockout blow against al-Qaeda and the Taliban: Operation Anaconda.

One hundred Australian SAS soldiers took part in the mission, the unit's largest battle since Vietnam. By the time it was over, the regiment would be credited with killing more than 300 enemy soldiers, mostly by directing air strikes at al-Qaeda positions.

In the days leading up to Anaconda, the SAS were inserted into the mountains surrounding the Shah-e-Kot Valley to feed intelligence about enemy troop locations. Working at high-altitude positions, usually about 3,500 metres above sea level, and from observation posts such as on the ridge called "the Whale's Back," they had a commanding view of any movement in the valley below.

The regiment's troopers were stationed at both ends of the valley for the battle. On March 2, an Australian patrol working with Green Beret Stanley Harriman's Cobra 72 team began moving into Shah-e-Kot as part of the force to drive al-Qaeda into the guns of coalition troops on the other side of the valley. As they advanced, they came under heavy fire from al-Qaeda positions and had to temporarily withdraw. The SAS, however, did manage to identify several al-Qaeda strongpoints before pulling back and those were later destroyed by air strikes.

On the same day, but on the other side of the valley, two SAS troopers tasked with conducting air support liaison duties for the U.S. Army's 10th Mountain Division, found themselves fighting for their lives after helicoptering into an enemy stronghold.

The two, a warrant officer and signaler, along with their 10th Mountain comrades, were inserted by MH-47 Chinook helicopter around 6:45 a.m. into the valley. The plan was for a total of 82 men, on board two Chinooks, to be dropped off and to take up blocking positions against the al-Qaeda who were supposedly being forced out of the valley by Afghan soldiers and coalition SOF. Instead, the SAS and 10th Mountain troops flew straight into an ambush.

Even as they were getting ready for the operation, the SAS men were given a warning that the mission may be more dangerous than first anticipated. Intelligence data had previously put the number of enemy expected to be operating in the area at about 100. Just as they were getting ready to board a Chinook helicopter, an American officer told them that the number was now estimated at about 500 and there was some talk that there could be more than 1,000 al-Qaeda in the

Shah-e-Kot area.

As their MH-47 touched down on a flat area around 2,900 metres above sea level, the SAS troopers knew almost immediately that they were in trouble. The two Chinooks came in on landing zones about 400 metres apart and men scrambled out of the aircraft to take up their positions as al-Qaeda gunfire echoed outside. There wasn't much cover; the terrain featured rocks and the odd juniper bush.

The troops had been told that their landing zone was well behind al-Qaeda positions. In fact, the location was in the middle of an enemy stronghold, complete with fortified bunkers and caves on each side of the landing zone. From these strongholds, al-Qaeda poured down gunfire. The Chinooks just barely managed to lift off, leaving the SAS and 10th Mountain troops rushing to take up defensive positions.

"There was no cover and 82 people were looking for some," the SAS warrant officer, known only as Clint P, later told a reporter with the Australian *Courier-Mail* newspaper. "We didn't understand what was out there."

The gunfire trailed off and then al-Qaeda opened up with mortars and more RPGs. The rounds were extremely accurate. U.S. Army Command Sergeant Frank Grippe, one of the senior 10th Mountain Division soldiers on the ground, figured that the enemy had been using the valley for so many years that they had already zeroed in their mortars on certain positions and were therefore able to walk in the rounds almost on top of the Americans and Australians.

It was indeed a familiar tactic. During the Russian war in Afghanistan in the 1980s, the Mujahedeen had wiped out a Soviet battalion under similar circumstances when the latter had landed on an open plateau. Three helicopters were shot down and the troops were caught in the crossfire from Mujahedeen positions on the slopes above them.

Grippe's hunch that the al-Qaeda mortars had already been zeroed in on specific positions would prove correct. During later sweeps through the valley, coalition troops discovered al-Qaeda bunkers where mortar base plates had been cemented into the ground, their firing positions lined up and ready to go.

As the mortar rounds increased in frequency, the soldiers began using their hands to dig out trenches for cover. The SAS signaler, Martin Wallace, used his knife to carve out a shallow pit in a dry creek bed. Clint P was able to dig out a hole big enough for about three men. In no time, about 30 soldiers were packed into his trench.

It was a highly dangerous situation. Just one mortar round landing in the middle of the group would cause large-scale casualties.

10th Mountain sniper teams kept some of the al-Qaeda pinned down while Wallace began feeding information back to the coalition command centre in Bagram about the group's desperate situation. The reply came back that air support was on the way.

Several Apache attack helicopters flew into the valley to strike at the al-Qaeda positions, but after making several runs at their targets they were forced back by a barrage of heavy machinegun fire and RPGs.

Next up was a B-52 bomber circling 6,500 metres above the besieged unit. Wallace could hear on his radio the pilot call out "bombs away" and he waited for the inevitable blast to rock the ground. Seventeen seconds later the bombs struck right on target, the incredible roar of their detonation echoing off the mountain walls.

But as soon as the debris settled, the al-Qaeda were up and shooting again. At times they even had the audacity to wave at the Americans and Australians.

Bombing from the air wasn't the only thing that proved ineffective. Because of the mountainous terrain, some radios the coalition troops were carrying were almost useless. Some U.S. soldiers who managed to work their way higher up the ridge used a runner to relay information to officers on the plateau below them.

Although the main enemy force occupied the eastern ridge of the valley, others were starting to appear on the western side too. Clint P could see about 26 al-Qaeda on those slopes moving into position. The two Australian SAS and their American comrades opened up on the group with their M4 carbines, killing several of them.

As the enemy continued firing RPG-7s and mortars, it was becoming apparent that the SAS and 10th Mountain soldiers were starting to run out of ammunition. The two SAS troopers believed that if they weren't airlifted out of the valley by morning, they would be overrun and killed.

As the sun started to drop behind the mountains, the al-Qaeda began maneuvering closer to the besieged troops and once again poured down an intense volume of fire, this time for about 25 minutes. "Probably the heaviest fighting was around last light, when they managed to dominate both of the ridge lines and launched a ground assault from the north," SAS signaler Wallace later recalled. "By the end of that they had set up a machinegun in the south so they had us surrounded. That was probably the scariest part of the whole day."

As the al-Qaeda amassed and began closing in, Wallace was on his radio requesting support from AC-130 gunships. Just as dusk was falling, several Spectres roared into the valley and began pounding the enemy positions. Rounds from the Spectres' howitzer and 40 mm cannons raked the al-Qaeda bunkers.

The AC-130s lived up to their feared reputation. Al-Qaeda and Taliban captured in other battles had referred to the aircraft as the "Spitting Witch" because of the flames that could be seen coming from its guns during nighttime operations. The aircraft would circle the battlefield pumping in rounds from its 40 mm cannon, its 105 mm howitzer and its mini-guns. The al-Qaeda and Taliban, who for the most part didn't have access to night-vision equipment, often couldn't see exactly where the fire was coming from. Some Spectre crews would later recount the surreal scene they viewed on their surveillance screens of a confused enemy scattering like ants as they were being gunned down by an unseen force in the darkness.

With the enemy pinned down, helicopters were able to land at about 8 p.m. and load up the wounded. Thirty-four Americans, most suffering shrapnel wounds from the mortar rounds, were taken out. Most of the injuries were to their legs and arms since the soldiers' body armor protected them from more serious wounds to the upper body. A little after midnight, a second wave of helicopters landed and the SAS men and their remaining American comrades piled on board.

Wallace would later receive the Australian military's third-highest decoration for bravery, the Medal for Gallantry, for his courage under fire.

Hours later in another location, one of Wallace's SAS comrades, Sergeant Matthew Bouillaut, engaged in a battle that would earn him a medal for his leadership under fire.

On the top of Takur Ghar mountain, the attempt to rescue U.S. Navy SEAL Neil Roberts, who had slipped off the back of a Chinook, had already begun. Members of Task Force 11 – SEALs and U.S. Air Force combat controllers – had been attacked while trying to recover their comrade and had been forced to retreat down the mountain. A Chinook carrying U.S. Army Rangers on a rescue mission for the beleaguered special operations troops had also been shot down and those men were fighting for their lives on Takur Ghar.

In an observation post on an adjoining mountain, Sergeant Bouillaut's SAS team spotted more al-Qaeda starting to climb Takur Ghar in an attempt to reinforce those already at the summit.

The five-man SAS patrol, Bravo Three, would become the lifeline for the besieged Rangers and prevent the larger al-Qaeda force from overrunning the American troops. Using their global positioning systems, the men, who were from the SAS Regiment's 1 Squadron, were able to determine positions of the al-Qaeda reinforcements on the slopes of the mountain and called in air strikes on the enemy formation. Over several hours, the SAS-directed bombs blasted into the al-Qaeda fighters, scattering their body parts into the crags and crevices of Takur

Ghar.

For his leadership of the patrol that afternoon, as well as the 11-day period that Bravo Three spent in the mountains during Anaconda, Sergeant Bouillaut received Australia's Distinguished Service Cross. "He displayed the highest level of commitment and dedication in commanding his patrol," the medal citation stated. "He made tactically sound assessments and decisions and displayed excellent leadership qualities under arduous conditions."

Operation Anaconda's commander, U.S. Army Major General Hagenbeck, would later acknowledge the SAS's key role in the mission, in particular that of Bouillaut's team in calling down air strikes to help the Rangers. "They were there, did what they needed to do – a supporting role in the first days but by the third day they were the main effort for the fight and they came through superbly," Hagenbeck said.

As Anaconda wound down, the SAS continued with their reconnaissance missions, cutting off the so-called "rat lines" – trails that the al-Qaeda and Taliban used to move in and out of the mountain region, as well as across the border to Pakistan.

Air strikes continued around Gardez, flushing out some enemy troops into ambushes set up by the SAS. At a March 15 briefing in Canberra, Australian military official Brigadier Paul Retter said the SAS had been successful in killing a number of fleeing al-Qaeda.

The number of enemy killed at the hands of the SAS in the closing days of Anaconda was estimated to be at least 10. The final days of the mission were spent uncovering ammunition dumps and arms caches such as one in the village of Oryakhail. In that raid, two SAS teams in long-range patrol vehicles moved into the village and seized a mobile anti-aircraft gun that had been hidden in a fort.

By late March 2002, a new contingent of 150 SAS was in Afghanistan, ready for action and to conduct what Australian officers termed "mopping up" operations. Their main mission was to conduct reconnaissance patrols in the southeastern part of the country, with particular attention to be paid to the area along the Pakistani border. Operation Mountain Lion, as it was called, would focus the hunt for al-Qaeda and the Taliban near the city of Khost, about 32 kilometres from the border.

On April 30, an SAS patrol conducting a surveillance mission on a suspected al-Qaeda building and nearby cave got into a brief firefight with four men. The SAS would later report they believed they killed or wounded two of the men, but the bodies were never found. A later sweep of the area by American troops revealed food, mortar rounds and several boxes of grenades.

But it was another firefight on May 16 involving two SAS patrols in the mountains about 40 kilometres north of Khost that would once again drive home some of the problems of the Afghan mission. The gun battle would raise questions about co-ordination of coalition special forces as well as allegations that Western governments desperate for results in Afghanistan were not above misleading the public about the success of some of their operations.

The battle started around noon on Thursday when an SAS patrol, Bravo One, conducting reconnaissance near the village of Chambagh, started taking fire from "enemy" forces. The four SAS soldiers dove for cover among the rocks as AK-47 bullets kicked up dirt and RPG-7 rounds flew over their heads. Trapped for several hours, they radioed for help from an SAS quick-response force based in Khost. As those operators, travelling in a column of heavily armed vehicles, rushed to the battle, they also came under attack from machinegun and recoilless rifle fire, about seven kilometres from Bravo One's position.

Shortly after 4 p.m., the Australians got some relief when two U.S. helicopter gunships circled the area and opened fire on enemy positions. The rocket and mini-gun barrage from the Americans allowed Bravo One to break contact with the enemy and move out on foot to link up with the quick-response SAS patrol. As they retreated, Bravo One was pursued by the enemy, who continued to fire mortars and RPGs at the Diggers.

Back in Australia, Brigadier Mike Hannan credited the U.S. air support for helping the SAS out of a tight jam. "The very effective close-air support was certainly instrumental in assisting the two groups to fight through the hostile forces and withdraw from this very dangerous situation," he noted.

Unknown to both SAS patrols, the combatants they were fighting were not al-Qaeda or Taliban. The special operations troops had been caught in the crossfire of a battle between two warring Afghan tribes, the Sabri and Mangal, who had been feuding over the last six decades. This time it was about which tribe had the right to cut timber on a mountain situated between their two villages.

The confused situation continued into the evening and, at about 10 p.m., three Spectre gunships and two Apache attack helicopters returned to the area, raking the mountain top with rocket and cannon fire for a solid 20 minutes. Ten Afghans of the Sabri tribe, most of them teenagers, were killed in the barrage. Two other Sabri were seriously injured. One Mangal tribe member was also wounded in the attack.

As details of the day's earlier fighting reached Bagram airbase, American officers ordered a contingent of Britain's Royal Marines to be put on full alert. After sitting around Afghanistan for months without firing a shot, the Marines were told they were finally going to war. Not knowing that the fighting involved feuding tribes, six Chinooks lifted off on Thursday night, taking the Marines into battle. Over the next several days, almost 1,000 British soldiers would be inserted into the area around Chambagh, along with a battery of 105 mm light guns to provide extra firepower.

British commander Brigadier Roger Lane told journalists his men were going after a "substantial" enemy force and would sweep and clear 50 square kilometres of the mountainous region near the position where the SAS had come under fire. Other coalition troops had taken up blocking positions to prevent the escape of enemy forces.

As Chinook helicopters continued to airlift the Royal Marines into position, village elders from Chambagh arrived in Khost for an emergency meeting with Governor Abdel Hakim Taniwal. There, they told Afghan government officials that they were under attack from coalition forces and pleaded with them to intervene with the Americans and call off the onslaught.

In an interview with the *Christian Science Monitor* newspaper, Haji Mohammed Hanif, an elder of the Sabri tribe, said he couldn't understand why his men had been rocketed and strafed. Weeks ago he had told U.S. special operations officers in Khost that he had armed men stationed on the mountain. The Sabri leader also said he had gone so far as to provide the Americans with maps showing their positions. "We are very disappointed, very unhappy," said Hanif. "We don't know why U.S. forces are killing us."

Afghan government officials were also frustrated. They said such mistakes could easily have been avoided if U.S. officers kept in contact with local commanders who knew what was going on in their areas.

But American military officials weren't buying into the Sabri's explanation. At a press conference in Bagram, U.S. spokesman Major Bryan Hilferty said the area was a well-known marshalling point for al-Qaeda and Taliban forces. "Usually people who fire at you are enemies," he responded when asked about the claims made by the Sabri.

But when journalists continued to push the issue, he acknowledged that it was "possible" the SAS troops had been caught in the middle of a local dispute.

Any suspicions that the special operations soldiers had been caught up in the middle of a deadly tribal argument weren't enough to stop Operation Condor, the code-name for the Royal Marines' mission in the Chambagh area. The Marines, loaded with heavy rucksacks full of equipment and ammunition, continued their sweep through the area.

Back in England, the news media carried contradictory reports about the mission. Brigadier Lane was quoted as saying his Marines were in pursuit of the enemy with orders to destroy them. An undisclosed number of al-Qaeda or Taliban fighters had been killed, he added.

Hours later, the British Ministry of Defense issued a clarification: the Marines had not killed anyone but coalition forces had caused casualties, a reference to the Apache helicopter and Spectre gunship attacks. In response to the flip-flop, opposition politicians in the House of Commons accused the government of hyping Operation Condor to justify the continued British presence in Afghanistan and to divert attention from the fact that Osama bin Laden had still not been caught, despite the large number of coalition soldiers scouring the countryside for the terrorist leader.

In any case, Operation Condor ended on May 23 without any shots being fired by the Royal Marines. They returned to their base, tired, dusty and somewhat cynical about their role in what many referred to as a wild goose chase.

It was left up to Royal Marine spokesman Lieutenant Colonel Ben Curry to explain to journalists the results of Operation Condor. "There was no contact with AQT (al-Qaeda and Taliban), but there were a number of small finds, including two 82 mm mortar tubes and a range of ammunition," he said, reading from a prepared statement. "An empty cave complex was also discovered and destroyed."

In early August, the Australian SAS rotated another contingent into Afghanistan to support the newly arrived U.S. 82nd Airborne Division. Over the months, contact with the enemy became less frequent as most of the al-Qaeda and Taliban forces decided to remain in their sanctuaries in the lawless tribal areas of Pakistan. By December 2002, the SAS considered its job done and headed home.

Its men returned to Australia to be awarded a number of medals. In addition, the Americans would award Bronze Stars to special force task group commander Lieutenant Colonel Peter "Gus" Gilmore for his service and to SAS Lieutenant Colonel Rowan Tink for his role in the planning and execution of Operation Anaconda.

While the Afghan mission was deemed a success, Australia's top special operations commander, Brigadier Duncan Lewis, warned that the battle against terrorism would continue in the shadows for a long time to come.

"The end of operations in Afghanistan should not be seen as the end of the fight against terrorism for Australia and our coalition partners," he said. "This is the type of operation which is likely to be fought over a long period of time, during which victories may be hard to see while reverses may be highly visible and public."

PREVIOUS PAGE, TOP: *Side view shot of the SAS 6X6 long-range patrol vehicle in Afghanistan showing both the .50 calibre and 7.62 mm general purpose machinegun. (COURTESY ADF)*

PREVIOUS PAGE, BOTTOM: *U.S. troops and Australian SAS meet in the Afghan desert. The Americans are using the Hummer while the SAS have their long-range patrol vehicles. (COURTESY ADF)*

THIS PAGE, TOP: *An Australian SAS trooper sits on the back ramp of a Chinook with a U.S. gunner. (COURTESY ADF)*

THIS PAGE, RIGHT: *An SAS trooper serving on Operation Slipper, Australia's contribution to the war in Afghanistan. He is wearing a radio earpiece and a front vest loaded with M4 magazines. (COURTESY ADF)*

TOP LEFT: *SAS motorcycles roar through an Afghan town during the hunt for al-Qaeda and Taliban. (COURTESY ADF)* / **TOP RIGHT:** *An SAS trooper takes a break during a mission while the snow-covered Afghan mountains loom in the distance. (COURTESY ADF)* / **ABOVE:** *Two SAS operators cross a small stream in their long-range patrol vehicle during Operation Slipper. (COURTESY ADF)*

OPPOSITE PAGE: *A JTF2 operator in Afghanistan in front of one of the unit's newly-acquired Hummer vehicles. Note the maple leaf near the front wheel. (COURTESY JTF2)*

DHC 02-163-12

Task Force Canada: JTF2 in Afghanistan

5 High above the Shah-e-Kot Valley, the Canadian military observation post was nestled in a flat spot in the snow, hidden behind a mountain ridge.

At times, al-Qaeda could be seen in the valley far below, bringing in supplies in four-wheel drive pickup trucks or on mules along the rat trails that headed in and out of Pakistan.

It would have been difficult for the enemy to spot the Joint Task Force Two operators. The commandos blended into their surroundings well, wearing white Canadian Army snow smocks and pants over their forest green uniforms. Their tent also had a unique camouflage pattern. It was mainly white, but speckled with brownish dots, making the shelter almost invisible from a distance. Snowshoes and packsacks were piled in a snowdrift.

As far as the eye could see, the majestic snow-covered Hindu Kush mountains rose to meet the deep blue Afghan sky. If there was one thing this special reconnaissance team had, it was a room with a view.

A few of the JTF2 operators were armed with C8 carbines, some of which had been outfitted with M203 grenade launchers, while at least one humped a C9, the Canadian version of the Belgian Minimi. Most of the soldiers had taped a field-dressing to the stock of their C8s for quick use in case they were wounded.

To get to their site, the operators had completed a dangerous tactical climb up icy gullies dotted with juniper bushes and fir trees. Still, the terrain seemed familiar, reminding some of the men of the foothills and mountains in the Canadian province of Alberta. Their rock climbing training also came in handy as they carefully navigated the boulder-covered slope.

Scattered throughout the mountains around them, hundreds of special operations soldiers conducting similar reconnaissance missions sat tight. Operation Anaconda had brought everyone from Green Berets and Delta Force to CIA operatives into action.

The weather, at times, had been miserable. Temperatures dipped to minus 20 Celsius at night and the mountain winds roared up some of the ravines shaking the JTF2 tent back and forth. In some places the snow was knee-deep.

Because of heavy sleet and snow, Anaconda commander Major General Hagenbeck had postponed "D-Day" by at least 24 hours. Visibility was too poor for the helicopters to bring in his main assault force.

But the JTF2 operators were in their element. While their SEAL and Australian SAS counterparts with Task Force K-BAR complained about the conditions, the weather in the mountains wasn't all that different from winters in the Ottawa Valley, where the men had their Canadian home base. This was the type of warfare JTF2 had trained for over the years, and now it was time to put their skills to use.

It probably seemed like an eternity to the operators, but it was only five months earlier that the Canadian government announced it would commit JTF2 to the American-led war on terror. The news had been greeted at the unit's base with excitement and enthusiasm as well as a bit of nervousness. For JTF2's commander, a Canadian Army Lieutenant Colonel, and his officers, the Afghanistan war was the unit's formal entrée into the world of special operations forces.

Although JTF2 considered itself a world-class counter-terrorism unit, it had never been tested on an actual SOF war mission. Originally formed in April 1993, JTF2 modeled itself on the British SAS structure but had also taken elements from Delta Force and the SEALs. Officially, its main role had been domestic counter-terrorism and hostage rescue. But the unit's officers had long debated transforming their organization into a full-fledged special forces operation, handling both covert "black" counter-terrorism missions and more traditional special forces wartime roles.

Based at the Dwyer Hill Training Centre in Ottawa, Ontario, JTF2 had originally started out with about 100 members. By the time of the September 11 attacks, its ranks had grown to 297.

Although the unit had been set up for domestic missions, such as dealing with

hijacked aircraft, JTF2 had also performed operations overseas. Its commandos had conducted specialized reconnaissance for Canadian units in the former Yugoslavia and made detailed plans for rescues in case the soldiers were taken hostage by the warring factions there.

In the summer of 1996, a JTF2 team was sent on Operation Stable, a mission to Haiti to advise on security arrangements for that country's president, René Préval, as well as to provide bodyguards for the Canadian commander on the island. In addition, JTF2 commandos taught tactical skills to the Haitian police SWAT unit. Under the Canadian tutelage, the Haitian tactical squad conducted a series of successful missions to search out arms caches held by anti-government forces throughout Port-au-Prince and the surrounding area.

Several months later, another JTF2 team provided close personal protection for a Canadian general and a diplomat who were trying to negotiate an end to the massive refugee crisis in Central Africa that had developed after the civil war in Zaire. And, in late 1996, JTF2 embarked on Operation Preserve, sending to Lima, Peru, the commanding officer of its 1 Squadron as well as the unit's operations officer to observe Peruvian SOF who were preparing to rescue hostages held by terrorists.

But it was the September 11 terrorist attacks on the U.S. that would change the unit forever. JTF2 officers quickly realized that the Afghanistan conflict would become a showcase for the unit's skills in front of special operations forces from other nations. As one officer later wrote, the Afghan war was a "significant milestone for Canadian SOF."

The unit selected some of its best men and officers for Task Force Canada, the name of JTF2's contribution to Afghanistan. Typical of that group was Mike Beaudette, who, according to published news reports, was a former member of the Canadian Airborne Regiment. The 42-year-old Beaudette had joined JTF2 as a major in 1998 when he passed the unit's special operations assaulter course. Later he was attached to JTF2's headquarters as deputy commanding officer.

About 40 members of the unit were on the ground in Afghanistan by late December 2001, moving in with other special forces operators at the Kandahar airport. The Canadians were immediately assigned to Navy SEAL Captain Robert Harward's Task Force K-BAR.

The airfield, just outside Kandahar, was a virtual beehive of special operations activity. Soldiers walked around the base, some in uniforms with their nametags removed. Most, however, decided to go for a blend of Afghan and civilian outdoor wear, one of the favorites being North Face fleece jackets.

Harward was keen to involve his foreign comrades in K-BAR's missions. At operations meetings, the specific missions were outlined and officers from JTF2 and the other foreign contingents each would list what they could offer.

JTF2 officers on the ground in Afghanistan were given much leeway in deciding whether to proceed with a specific operation. It was only on the more complicated and dangerous tasks that a "targeting" committee made up of senior Canadian Forces leaders back in Ottawa would review mission details and give its approval for the operation to proceed.

Inserting JTF2 into K-BAR was also made easier by the fact that the Canadian commandos had carefully crafted alliances with some of the world's top counterterrorism and SOF units, many of whom were now operating in Afghanistan.

JTF2 had a long-standing link with the SEALs, in particular SEAL Team Six. As early as July 1996, JTF2's 3 Squadron had hosted SEAL snipers at Canadian Forces Base Petawawa, Ontario, while 1 Squadron had trained with another group of SEALs at Meaford, Ontario. Those exchanges were soon followed by a visit by JTF2 members to SEAL Team Six's facilities for "Exercise Cable Lunge." The training scenario was a "no notice" deployment which sent the Canadian operators to brush up on their shooting skills at the navy unit's "killing house" in Virginia Beach, Virginia.

Exchanges had also been going on with the British SAS. In April 1996, the commanding officer and the Regimental Sergeant Major from the British unit toured JTF2's Dwyer Hill base and came away suitably impressed. That was followed by reciprocal visits to the SAS base in Hereford, England.

JTF2 also didn't neglect its British counterparts in naval special operations. In 1999, there was a small unit exchange with the SBS for Operation Hibernia as JTF2's dive team joined forces with the famed British commandos for practice raids on a Hibernia oil rig over an eight-day period in August.

The Canadians also continued JTF2's excellent working relationship with Delta Force when they hosted that unit's snipers in April 1998.

The year preceding the September 11 attacks was one of the busiest for JTF2 in terms of training with foreign counterparts. For example, JTF2's commanding officer and other operators visited Delta Force for two days in March. As well, six members of the New Zealand SAS were at Dwyer Hill for several months starting in August. And, in November, members of the British SAS were back at Dwyer Hill for training with 2 Squadron, while a few days after that a group of SBS hooked up with 4 Squadron for a maritime counter-terrorism training mission dubbed Exercise Hydra.

All this work in forging relationships with the world's premier counter-terrorism and SOF units would pay dividends for JTF2 in Afghanistan. Even though the unit was virtually unknown to the Canadian public, its commandos had gained a solid reputation among other operators.

Sensitive site exploitation missions were some of the first that JTF2 conducted in conjunction with other K-BAR troops. Several of the Canadians were sent to help with the exploration of al-Qaeda's Khawar complex after it became apparent that the large number of tunnels and caves was too much for the SEALs to search by themselves. For some of the site exploitation missions, a few JTF2 operators armed with C9 machineguns pared down the size of the weapons by removing the stocks. This reduced the ability to accurately aim the C9, essentially transforming it into a spray gun for close-range firing in caves and rooms of al-Qaeda and Taliban compounds. In light of the nature of some of K-BAR's operations, JTF2's decision to go for a shorter and more maneuverable weapon was a smart one.

On some missions, one or two JTF2 operators were assigned to work with four or five American commandos, along with an Afghan guide and interpreter. In later missions, JTF2 worked in four- or six-man teams on its own.

The unit had set up its own compound at the Kandahar airfield, outfitted with an outdoor weight room and comfortable bunks. Strict rules were in place regarding the news media: there were to be no photos taken of special operations personnel and no discussions about their missions in articles or broadcasts. Canadian journalists watched JTF2 operators with their American comrades driving through the dust-filled streets of Kandahar in the back of Toyota pickup trucks, but because of the rules, the Canadian public was kept totally in the dark about the unit's activities and successes.

All that changed following a January 20, 2002 raid on an al-Qaeda complex during which JTF2 grabbed several prisoners.

Associated Press photographer Dario Lopez-Mills was at the Kandahar airbase when he saw a large U.S. Air Force MH-53 Pave Low helicopter come in for a landing. Military officials told him to leave the tarmac and go inside a building, which he did. But as Lopez-Mills looked back through the door he could see three soldiers coming off the ramp of the helicopter, their hands clenched roughly around the backs of the necks of three men dressed in white. Lopez-Mills lifted his camera and quickly snapped one photograph.

On January 22, the picture was featured in newspapers around the world, with the caption identifying the masked commandos as U.S. troops coming back from a successful mission. It was the type of photo that best captured the shadow war being waged by special operations soldiers in Afghanistan.

A little more than a week later, on January 29, when it became known that the commandos in the photo were JTF2 operators and not Americans, a political firestorm erupted in Canada.

The issue about what to do with al-Qaeda and Taliban prisoners had already been the subject of heated debate in the country. Defence Minister Art Eggleton had announced that any captives taken by Canadian soldiers would be turned over to the U.S. government, which was on record as stating that they would be treated as unlawful combatants. There would be no Geneva Convention for the Taliban or al-Qaeda, although American officials said they would be treated humanely.

In Canada, there was a great deal of soul-searching and debate among politicians as to whether international law allowed the country's military forces to turn over prisoners to a nation that was not going to abide by the Geneva Convention for those individuals. Even members of the ruling Liberal government voiced concern that such actions undercut the long-standing rules of war.

On January 28, Prime Minister Jean Chrétien, apparently unaware that JTF2 had already taken prisoners, tried to defuse the controversy. According to the prime minister, the whole issue was hypothetical since Canadian troops had not yet been involved in taking any captives.

But a day later, in a bizarre turn of events, Eggleton told the Prime Minister that the unit had indeed seized a group of prisoners on its January 20 mission. Later he blurted out to a group of journalists that the commando unit had grabbed suspected terrorists, making a point of singling out the Lopez-Mills photograph. "Did you notice the fact that they had forest-green uniforms?" Eggleton said of the three soldiers in the picture. "Well, they were Canadian JTF2."

Eggleton's stunning admission prompted a House of Commons inquiry to investigate whether he had deliberately misled Parliament and kept the prime minister in the dark. It would later be revealed that Eggleton already had been briefed by military officials on January 21 about the successful capture of the prisoners but had failed to pass that information along to other members of Cabinet, including Prime Minister Chrétien.

At National Defence headquarters in Ottawa, there also seemed to have been a disconnect, along with some naiveté about the situation on the ground. For example, even though Canada was, arguably, violating the Geneva Convention by turning the prisoners over to the U.S., senior officers still expected the Taliban and al-Qaeda to treat any captured Canadians by those rules. Further, in a report prepared on the prisoner issue, senior defence officials outlined how enemy forces would be required to allow Red Cross officials to visit any Canadian POWs.

JTF2 operators in Afghanistan, however, were under no such illusions. This was a war where no quarter was given or expected. JTF2 soldiers knew that, if caught, they would be executed outright by the Taliban and al-Qaeda or tortured until death.

As Eggleton concentrated on trying to ride out the POW controversy (he would later acknowledge he made a mistake by not promptly informing the prime minister about the capture of al-Qaeda suspects), officers at the JTF2 base at Dwyer Hill focused on acquiring additional equipment for the Afghanistan mission.

New night vision goggles would be sent to the operators, with the older ones returned to Dwyer Hill for modifications, such as installing more robust metal parts instead of plastic ones. Radios would also have to be improved, with some being sent back to the manufacturer for software upgrades.

But the main equipment priority for JTF2 in Afghanistan was to get a reliable vehicle so the operators could conduct more wide-ranging missions.

Before the war, JTF2 had been using commercial SUVs and a heavily modified truck already in the Canadian Forces inventory, known as the light support vehicle wheeled (LSVW). Some of the unit's LSVWs had been equipped with mounts designed for heavy weapons on the front and rear. The vehicles' cabs had been removed and global positioning systems installed. The advantage of the light trucks was that they could be transported by CC-130 Hercules aircraft. But the LSVWs were renowned in the Canadian Forces for their poor reliability and performance. The trucks had been plagued by engine fires and were considered by operators to be under-powered – two rather troublesome features in a combat zone such as Afghanistan.

While assigned to K-BAR, JTF2 had access to four-wheel drive commercial pickup trucks. But what its operators really needed was their own fleet of Up-Armored Hummers or HMMWVs (High Mobility Multi-purpose Wheeled Vehicle). Eggleton signed off on the $641,000 (U.S.) purchase at the end of January, and by April the first of the vehicles were in theatre in Canadian hands.

The Hummers were outfitted with .50 calibre machineguns as well as a top rack that could hold at least three ammunition boxes. A small maple leaf decal was put on the front of the vehicles to designate their Canadian pedigree.

Selecting the M1114 Up-Armored Hummer made sense from an operational point of view. The vehicle provided JTF2 with good protection and mobility. With more than 150,000 Hummers in the American military inventory, and the U.S. being Canada's main ally, it was also a good logistical fit. Not only would the vehicles make JTF2 interoperable with their U.S. counterparts, but spare parts would be readily available on overseas missions from the Americans.

. Interoperability with allies was a major issue for U.S. special operations forces and one that found JTF2 as well as the other K-BAR coalition SOF lacking. JTF2 used essentially the same weapons, Sabre radios, night vision goggles and other personal gear as the U.S. operators. What was lacking, however, were the larger pieces of equipment such as fixed-wing aircraft and helicopters. K-BAR's Captain Harward would later note that all of the coalition operators assigned to him were top-rate, but that the lack of interoperability of their equipment posed problems in conducting the war and could possibly affect future combat situations where allies were used.

"They are all great shooters ... but they can't integrate their systems ... they can't bring their gunships," he told *National Defense* magazine.

That was at least one area JTF2 didn't have to worry about since Canada didn't possess any gunships.

Since the unit was integrated into K-BAR, it would have access to the U.S. Air Force's MH-53 Pave Lows and the U.S. Marines Corps' CH-53 helicopters, both with the required space for equipment and the range and power needed to operate in the mountains.

Equipment issues aside, the JTF2 operators in Kandahar concentrated on preparing for new missions. The next target for special operations troops would be Kandahar's Mirwais hospital where six al-Qaeda fighters, armed with pistols and grenades, had been barricaded for almost two months.

The six had been left behind at the hospital after the Taliban abandoned the city on December 7 as American SOF and their Afghan allies advanced. The al-Qaeda, most believed to be from Yemen and the Sudan, were threatening to blow themselves up with grenades if anyone but doctors entered their ward. They had used beds and mattresses to barricade the windows and doors of a group of rooms in the hospital. They also had stockpiled food and water.

Afghan troops, allied with the U.S. and trained by Green Berets, had surrounded the hospital and over the weeks negotiations had continued in an effort to convince the al-Qaeda to surrender. The men, however, seemed determined to martyr themselves. One doctor who examined a wounded al-Qaeda saw that the man had taped a grenade to his thigh in case he needed to commit suicide. Another had managed to slip out of the hospital but blew himself up with a grenade after Afghan troops tried to take him prisoner.

After issuing a series of ultimatums, all of which were ignored, planning began in earnest for an assault on the hospital.

Ghulam Mohammed, the head nurse at Mirwais, told a journalist with Canada's *Globe and Mail* newspaper that before the raid he was questioned by two special operations soldiers, one American, one Canadian. The men had requested information about the al-Qaeda holdouts and details about the physical layout of the hospital.

Preparations for the raid had already begun at a former home of Taliban leader Mullah Muhammed Omar. There, special operations troops, led by Green Beret Captain Matthew Peaks, began training about two dozen Afghan soldiers in how to storm the Mirwais. A mockup of the ward where the al-Qaeda were barricaded was built and the Afghans were put through "room clearing" drills. The plan called for the special operations troops to blow a hole in a wall of the al-Qaeda ward and send in the Afghan soldiers. Those men would have to move down a corridor towards the barricaded al-Qaeda; it was imperative that as they passed by various rooms they ensured each was cleared of the enemy. Such a tactic would prevent al-Qaeda from suddenly appearing from behind and opening fire.

In the early morning hours of January 28, the special operations troops and their Afghan counterparts conducted one final practice run on how the raid would unfold. By first light, they were ready to go. Roads leading to the hospital were blocked off and SOF snipers were positioned on rooftops overlooking the facility. The breaching team slipped inside the hospital and went about rigging their explosive charges to blast open a hole in the wall of the al-Qaeda ward.

Green Beret Major Christopher Miller, sitting in a command post about 50 metres away from the hospital, gave the go-ahead for the mission.

Inside, on Captain Peaks' command, one of the SOF ignited the breaching charge and the blast sent brick and plaster flying, as the explosion tore a hole into the wall. The Afghans rushed through the opening, tossing grenades as they moved down the corridor towards the barricaded al-Qaeda. Explosions echoed in the hospital as the Afghans, in their eagerness to attack, rushed forward and were promptly injured by shrapnel from their own grenades. From outside, other special operations troops tossed more grenades into the second-floor al-Qaeda ward and soon smoke was seen billowing from the hospital.

As their room started filling with acrid smoke, two of the al-Qaeda moved near a window to try to get fresh air but were immediately spotted by the special operations snipers. "Take them out," came the command over the radio and two snipers opened fire, killing the men.

After another failed attempt by the Afghans to enter the al-Qaeda ward, it was decided that SOF would lead the way for a final assault. In preparation, the special operations soldiers put on Kevlar helmets and body armor. The Afghans, now with six wounded, did the same.

Three SOF and three Afghans crept down the corridor towards the al-Qaeda stronghold and, when they were close enough, tossed in several grenades. Inside the room, the al-Qaeda quickly picked up a few of the bombs and tossed them back. A blast ripped through the corridor as the assault force tried to take cover. The Americans threw in more grenades before Afghan soldier Abdullah Lalai and one of the SOF soldiers stormed into the barricaded room. Lalai shot one of the enemy in the head and another in the chest. The American operator dove onto the floor firing his machinegun under a bed where two more al-Qaeda were hidden. All six of the terrorist hold-outs died in the raid.

The hospital ward had turned into a charnel house. Body parts from the dead al-Qaeda were strewn about and the floor was covered by large pools of blood.

U.S. military officials both in Kandahar and Washington credited Afghan troops for planning and running the successful operation. Major Miller said Afghan troops took the lead in storming the hospital and that American soldiers were there only to advise and assist.

"I think it's fair to say that because it was Afghan-led, that the Afghans properly get the credit for having brought this to a conclusion," Rear Admiral John Stufflebeem told journalists at a Pentagon briefing.

There was no mention of the Canadian involvement, but the hospital's head nurse, Ghulam Mohammed, said he could see that some of the special operations soldiers involved in the fighting were wearing the distinctive Canadian forest-green camouflage uniforms. A U.S. special forces officer also told *Globe and Mail* reporter Mark MacKinnon that JTF2 was indeed involved in the attack.

Back in Ottawa, Defence Minister Eggleton was asked about the hospital raid during his appearance before the Parliamentary committee looking into the POW affair, but he declined to answer, citing the need for JTF2's operational security.

The reluctance to discuss any details whatsoever about JTF2 was an ongoing dilemma for the Canadian Forces. On one hand, JTF2 considered itself a Tier One counter-terrorism unit, similar to the British SAS and Delta Force. That meant, according to Canadian military officials, that no photographs or details of JTF2 operations, equipment, personnel and training were ever to be made public. At the same time, JTF2 officers were keen to build up a "mystique" about their organization and the lack of solid information about the commando unit helped fuel that.

On the other hand, there were several problems with this approach. JTF2 still had to recruit from the ranks of the Canadian military, a near impossible task if all information on the unit was to be kept top secret. As well, senior military leaders and politicians appeared confused about what the actual policy was; the Defence department had already released photographs and videotape of JTF2 in training and had put up biographical information about some of its officers on the official

Canadian Forces website. Furthermore, Eggleton had already publicly revealed some of JTF2's missions. In addition to the January 20 raid which was done in conjunction with U.S. SOF, the minister also acknowledged the unit had been present on an earlier mission in January in which coalition forces took prisoners. In that case, JTF2 did not actually take custody of the captured enemy.

Then there were the hundreds of journalists in Afghanistan and, as the POW photo controversy showed, it was extremely difficult to control them.

Not that the Canadian Forces didn't try. The POW photo had so spooked the senior leadership that a military public affairs officer phoned at least two major media outlets – Southam News and Canadian Press – trying to get a commitment that they would not publish any photos their journalists might obtain of JTF2 in Afghanistan. The news outlets refused to make that commitment.

At the Kandahar airfield, Canadian military public affairs officers threatened any journalists with expulsion from the installation if they dared to write about special forces operating from the base. Some reporters were even told not to look in the direction of the JTF2 compound as they walked by.

It didn't take long for the military public affairs officers to follow through on their threats. On February 12, a reporter with the *Toronto Star* newspaper was banned from the base after he described the number of guard towers around the prisoner detention centre and mentioned that JTF2 had conducted night operations.

While the military public affairs officers did their best to keep the Canadian public in the dark about JTF2's Afghan mission, there was one person they couldn't control – U.S. CENTCOM commander General Tommy Franks.

No one had told the general about JTF2's supposedly iron-clad security policy. So, during a live televised news conference on March 4, 2002, Franks named Canadian and other foreign special forces involved in Operation Anaconda in the Shah-e-Kot Valley. The commandos, he said, had set up their observation posts about a week before the mission began.

At National Defence headquarters in Ottawa, military officials were in a panic. Canadian officers continued to refuse to say whether JTF2 was involved in Anaconda, even though Franks had announced it at the press conference being carried on almost every North American TV station. The officers even went as far as refusing to give out the phone number for the public affairs branch at CENTCOM's headquarters in Tampa, Florida.

Only later in the day would Canadian officers reluctantly confirm that Franks was right, but still they refused to give any additional details.

For the Anaconda mission, K-BAR's Harward put into the field 72 special reconnaissance teams, which included JTF2 as well as operators from Germany, Norway, Denmark and Australia. Recognized as specialists in winter warfare, the Canadians were a natural selection to operate in the snow-covered mountains. An estimated 20 to 25 JTF2 took part in Operation Anaconda with most of their surveillance missions taking place at altitudes above 3,000 metres.

After their successful mission was over, JTF2 operators continued to work closely with their U.S. counterparts, conducting long-range patrols and searches of sensitive sites such as caves and Taliban compounds. These raids, which sometimes yielded documents and weapons, also at times created bad will among the Afghans, who accused the coalition special operations troops of being heavy-handed in their tactics.

One such controversial mission took place on May 24 at the village of Band Taimore, 80 kilometers northwest of Kandahar.

A strike force of U.S. special forces, JTF2 and members of the U.S. 101st Airborne landed by helicopter near the village around 1 a.m. More than 150 troops – half paratroopers, the other half SOF – rushed from several Chinook choppers and quickly set up a perimeter around the collection of huts that were surrounded by a mud brick wall.

U.S. intelligence had identified the village as a potential hotbed of Taliban and al-Qaeda supporters.

A breaching team used explosives to blow several two-metre holes in the wall and JTF2 commandos and U.S. paratroopers went from building to building rounding up the terrified Afghans.

The soldiers kicked in the flimsy doors of the huts and tossed stun grenades into the homes. Men, women and children were dragged off the floor mats they had been sleeping on and were handcuffed with plastic restraints. JTF2 operators put chemical glow sticks around the necks of some of those they had captured, a method often used in training at Dwyer Hill to designate a suspected enemy.

The eight-hour mission would be steeped in controversy with allegations that innocent people had been injured or killed. One man, 70-year-old Haji Bajet, who was to be the village's representative to the loya jirga, or Afghan grand council in Kabul, died of his injuries while in American custody. Two others, including a man who had been sleeping in a car in the middle of the compound, were wounded after soldiers alleged they opened fire on the assault force or had tried to flee. The white stationwagon the man had been sleeping in was riddled with more than 40 machinegun bullets.

Inside the village mosque, blood stains and a bullet hole marked the concrete floor where one person had been shot and then dragged away by coalition troops.

After the soldiers left, the villagers discovered that a three-year-old girl had been killed when, during the confusion and panic of the raid, she had tumbled into a well. Her body was found at the bottom of the 12-metre shaft, partially covered in water. Another girl was hurt later in the day after a discarded stun grenade she found and played with exploded, injuring her foot.

The commandos took 55 of the male villagers into custody, including a 12-year-old boy, on the belief they might be associated with the Taliban or al-Qaeda. Later, at a U.S. military processing centre, the Afghans were stripped, photographed and interrogated.

This time, government officials in Ottawa decided to acknowledge the raid and JTF2's contribution in a roundabout way. Although not specifically mentioning the unit, they did note that Canadian troops not assigned to the conventional force battlegroup in Afghanistan had taken part in the mission.

"This is one more example of the ongoing contribution to the coalition efforts to fight al-Qaeda and the Taliban," said Defence Minister Eggleton. "Canadian troops continue to take an active role in the elimination of the terrorist threat."

The raid, however, seemed only to turn the villagers against the coalition. Since the village was close enough to Kandahar, journalists were on the scene shortly after the strike force had left and villagers told angry tales of humiliation at the hands of the troops. The men not taken captive alleged they were slapped and punched by the soldiers and the women were shoved and tied up, highly offensive behavior in any society but more so in the Islamic culture. One woman said a soldier used a turban to gag her while another alleged her seven-year-old daughter had her hands bound with plastic restraints.

Military debris was scattered throughout the compound with at least eight used stun grenades piled on the ground as well as a C9 plastic machinegun magazine with a belt of live ammunition. At some point, some of the commandos must have had time for a bite to eat, as a package of Canadian-made rations was discarded at the scene.

After the raid, 150 Afghans from the area around the village went to Kandahar to protest to the governor about the mission and demand that the detainees be released. Some of them warned that the heavy-handed special operations missions were driving the local population to support the Taliban.

Six days after the raid, the Americans released 50 of the captives, including the boy, after it was determined they had no affiliation with enemy forces. Five of the men were kept in custody for further interrogation, with one being described by U.S. Defense Secretary Rumsfeld as a Taliban official "below the senior level."

The villagers who were released had been subjected to several days of interrogation at a secret site. After they were freed, they were transported by Chinook to a soccer field at Kandahar and told to make their own way home. Several complained they had been punished for talking to each other and had been forced to crouch in the hot sun for several hours.

Each of the villagers was shown pictures of high-ranking Taliban and al-Qaeda officials and asked if they knew any of them. Among the photos displayed to the men were bin Laden, Taliban leader Mullah Mohammed Omar, and two of his deputies. Before they were released, the Americans told the men about the September 11 attacks and tried to explain why they had come to Afghanistan.

Pentagon officials maintained that the special operations soldiers acted professionally. Force was only used when several people in the compound tried to leave the area and another opened fire on the assault teams. "We went in there to see what was going on, and people don't shoot at you [as] a normal course of action," Lieutenant Colonel Jim Yonts, a spokesman for the U.S. Central Command, told journalists.

But what Yonts didn't say was that in Afghanistan, where robberies and killings by bandits and attacks by rival tribes were common, it would be a normal course of action to open fire on unknown intruders, especially those who had just blasted their way into a village's compound.

Canadian military public affairs officers went even further. They said that since there was no photographic evidence that any villagers had been killed or injured, there was no proof the claims made by the Afghans were true. (A crew from the Canadian Broadcasting Corporation had, in fact, videotaped the little girl injured by the discarded stun grenade and several days after the raid the U.S. military did return the body of Haji Bajet to the villagers. The Americans also acknowledged that Bajet was killed as a result of the raid and two other Afghans had been wounded.)

The raid, and the continuing hard-line tactic of swooping down on Afghan villages and snatching prisoners, raised more than a few eyebrows among British officers. They believed that coalition forces should be waging a "hearts and minds" campaign to win over the Afghans. Instead of midnight assaults on compounds, they suggested that special operations forces create small bases at sites across Afghanistan, giving the coalition an ongoing presence in a particular area. That would allow operators to work with the locals and build trust, making it easier to determine who was the enemy. The concern among the British was that the ongoing raids were giving Afghans the impression that the coalition was just another invading foreign army that had no respect for the country's culture or religion. However, there would be no change in tactics.

Back at Dwyer Hill, the decision was made to rotate a new contingent of operators into Afghanistan, not only to give those overseas a well-deserved rest but to allow a new group to gain combat experience. The first rotation of JTF2 commandos returned at the end of May 2002 via a Canadian staging base in the United Arab Emirates. Back home, they were personally greeted and thanked for their contribution to the war on terror by new Defence Minister John McCallum.

It was McCallum's first day on the job and JTF2 officers were gratified that the minister met the men, a true sign of government acceptance of the unit despite the controversy over the POW affair. In fact, while Eggleton had positive things to say about JTF2, McCallum would become one of their most effusive supporters. In late July 2002, JTF2 provided close personal protection for the defence minister while he toured Canadian operations at Kandahar. He would later write that "under the tender and extraordinarily competent care of our special forces, known as JTF2, our group felt perfectly safe at every moment of our stay in Afghanistan."

It was likely the first time that JTF2 had ever been called tender.

By the summer of 2002, the role for special operations forces in Afghanistan had grown smaller as most of the day-to-day operations were being taken over by conventional U.S. troops. McCallum announced that JTF2 would remain in Afghanistan for at least several months but that any commitment of the unit after that would depend on the situation on the ground. "There (is) a decreasing number of bad guys around," he noted.

On October 2, 2002, JTF2 officers discussed with senior staff about returning the contingent to Dwyer Hill. There were indeed fewer missions and the men were not being used to their full potential. It was time for "an operational pause," according to Canadian Chief of the Defence Staff General Ray Henault.

McCallum, as usual, had nothing but praise for Task Force Canada. "I think Canadians would like to know that this group is capable of doing amazing things.... I can say that in terms of their ability to do amazing physical acts, it is quite something," he told journalists.

Not that the Canadian military had any intention of letting the public learn any details about those amazing abilities. In fact, it had worked to quash a suggestion by McCallum that more information about the unit's activities in Afghanistan be made public. In December, senior military officers also moved to close down what they saw as a potential security problem for the commandos: revealing any honors bestowed upon JTF2 operators.

Unlike other militaries, the Canadian Forces tended to hide the achievements of its troops, particularly those involved in combat operations. Over the years, the government and Defence department had keenly promoted the idea that the Canadian Forces was an organization of peacekeepers. Killing the enemy, it seemed, was not what soldiers, at least those in Canadian uniform, did. So when Canadian troops went into battle in 1993 at the Medak Pocket in the former Yugoslavia, killing more than a dozen Croatian soldiers during one fierce firefight, Defence department officials ensured that details were kept from the public for years.

The same would hold true for JTF2. Senior officers came up with a plan to ensure that details about the heroism of individual commandos or any official recognition for the unit's exploits didn't leak out. That would be done in two ways. First, published information about the awarding of medals for JTF2 operators would be "sanitized" to such a degree that the public would never realize what a particular soldier had done to deserve the honor, nor that the individual was from the counter-terrorism unit. Any announcement of a medal would be bland enough not to generate interest among the news media or the public. The second method was the recommendation that the country's Governor General seal any medal citation forever from public view.

For the officers, such efforts were deemed necessary to protect the identity of the JTF2 operators as well as their families. And, again, the policy was in keeping with the strict security regime adopted by some units, such as the British SAS.

Senior Canadian military leaders rejected outright the route taken by other special operations units with Tier One capabilities, such as the Australian SAS. In the Afghanistan war, that unit allowed the identification of some selected troopers and details of their heroism and medal citations to be made public. In addition, senior SAS officers who were awarded Bronze Stars by the U.S. were named. In other incidents, the unit released details of the medal and a soldier's heroic actions but didn't identify him by name. In such cases, the Australians identified individuals by the pseudonym "Trooper X."

JTF2 operators did earn Canadian medals and honors for their fine work in Afghanistan, but the public has never been told about these nor any details of the operators' accomplishments on almost 36 missions. The only official Canadian word on their Afghanistan activities came in a JTF2 information video released in 2003. In that video, Deputy Chief of the Defence Staff Vice Admiral Greg Maddison states that JTF2 "captured enemy personnel, equipment and material of significant intelligence value and hampered the enemy's ability to conduct operations against us and our coalition partners."

No other details were given.

In fact, after K-BAR commanders, in a series of open briefings and presentations, told American politicians and journalists about the excellent work JTF2 and other foreign units had done in Afghanistan, the word was politely passed down to the U.S. SOF community that the Canadians were not to be mentioned in public ever again.

LEFT: *These JTF2 operators are outfitted with a combination of different uniforms. The man on the left is wearing a Canadian CADPAT while the one on the right appears to have U.S.-issue camouflage. Both have Canadian Army snow smocks.* (COURTESY JTF2)

BELOW: *Climbing up a snow-covered gully, two JTF2 operators take a few moments to catch their breath. The unit conducted high-altitude reconnaissance missions for Task Force K-BAR.* (COURTESY JTF2)

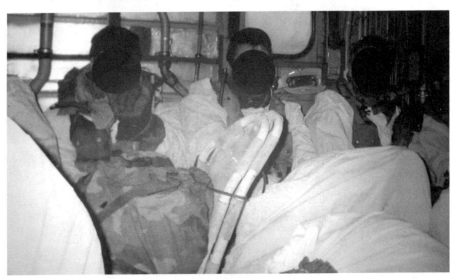

THIS PAGE (TOP): *Two JTF2 soldiers armed with C8 carbines on a mission in Afghanistan. The operator on the right also has an M203 grenade launcher outfitted on his weapon. (COURTESY JTF2)* ***(BOTTOM):*** *A JTF2 team on board an American military helicopter as it transports them on a mission to establish a high-altitude observation post. Note the snowshoes in front of the operators. (COURTESY JTF2)*

OPPOSITE PAGE: *Chechen terrorist Movsar Barayev, who led the hostage-taking at the Dubrovka theatre in Moscow, was killed in a firefight when Alpha commandos stormed the building. (COURTESY NTV)*

Siege in Moscow

6 By the time the Australian SAS and Canadian JTF2 were leaving Afghanistan in the fall of 2002, the war had turned into a series of sporadic skirmishes between remaining special operations forces and Taliban and al-Qaeda fighters holed up along the country's mountainous border with Pakistan.

In the hunt for Osama bin Laden, the trail had all but gone cold. U.S. special forces were still searching the lawless frontier along the Afghan-Pakistan border in the hopes of capturing the elusive Arab, but their efforts had been fruitless. Bin Laden and his men had gone underground, likely hiding in one of Pakistan's large cities.

In other parts of the world, however, al-Qaeda-associated terrorists were making their presence known. On the resort island of Bali, for instance, al-Qaeda sympathizers detonated a massive bomb in October, killing 202 people, many of them Australian tourists. In the Philippines, the Abu Sayuf group killed dozens of civilians in a series of bombings and attacked government patrols on a regular basis.

And in Moscow, Chechen terrorists would embark on an audacious large-scale hostage-taking just five kilometres from the Kremlin that would tax the skills of special operations troops to the limit. By the time it was over, some Russians would label the tragedy a smaller-scale version of the September 11 attacks on the U.S.

Though different in scope and history, the American-led war on terror and

Russia's war against Chechen terrorists shared similarities. For starters, they both involved extremists who had struck at the very heart of each nation. In fact, the Kremlin saw the Chechens' actions as part of a larger plan by al-Qaeda and associated groups to impose a global Islamic rule. And for both nations, dealing with these terrorists meant bringing in the very particular skills of their special operations forces.

The hostage-taking began on a crisp Wednesday evening on October 23, 2002, when three vans pulled up outside Moscow's Dubrovka theatre centre, where the hit Broadway-style musical, *Nord-Ost*, was being performed. In the vans were 41 Chechens, 19 of them women, armed with an assortment of grenades, AK-47s and pistols.

The black-clad women and men outfitted in camouflage combat fatigues rushed into the theatre's spacious auditorium, some taking positions near the exits while others stormed the stage. At first, they were hardly noticed. Many of the 800 people in the audience thought the armed intruders were cast members since *Nord-Ost* featured actors in military uniforms dancing and singing.

It was only after the Chechens started firing their pistols into the air that panic erupted. Audience members started yelling and screaming, with some trying to push their way out of the auditorium. Escape, however, was impossible, as armed terrorists blocked the exits.

As the sound of gunfire echoed through the auditorium, some of the play's production staff who had been working in the upper floors of the Dubrovka quickly realized something terrible was happening downstairs. To escape, they made a rope out of costumes and slid down the outside of the building from the theatre's third floor. One producer slipped off the makeshift rope and plummeted to the ground, breaking his leg. Still, unlike those inside, at least he was free.

After gaining control of the Dubrovka, the Chechens got down to business. Thirty explosive charges were rigged at points throughout the theatre. Two large bombs, each containing 45 to 50 kilograms of explosives, were set up near the middle of the auditorium.

The group's leader, Movsar Barayev, quickly made contact with Russian officials and set out his demands. He wanted an immediate end to the war in Chechnya and the withdrawal of all Russian troops from the breakaway republic within seven days.

The Muslim republic, located in the oil-wealthy North Caucus region, had declared its independence in 1991 when the Soviet Union began to split up. The Chechens had a long history of intense hatred for their Russian conquerors who had ruled their land for more than 100 years. During the Second World War, the

Chechens had welcomed and supported invading Nazi forces who briefly occupied the region and pushed out the Russians. But after Germany lost the war, the
Russians took their revenge and hundreds of thousands of Chechens were mur-
dered in ethnic cleansing and deportations to Gulags which lasted until the late
1950s.

In an attempt to crush the Chechen separatist movement, Russia invaded the
breakaway republic in 1994, but quickly became bogged down in a brutal war.
Over the next 20 months, Chechen cities were bombed into rubble and an estimated 80,000 to 100,000 civilians were killed in the fighting.

Even so, the Russian military proved to be no match for the well-organized
and aggressive Chechen guerrillas. Russia withdrew its forces in 1996 but once
again invaded the country after a wave of terrorist bombings three years later was
blamed on the Chechens. The bombings flattened apartments and buildings in
several Russian cities, and killed more than 300 civilians.

The new war was at a stalemate. More than 200,000 Chechens had fled their
country and were living in refugee camps scattered throughout the region. Russian troops controlled much of Chechnya but were still losing 10 to 15 soldiers a
week in hit-and-run rebel attacks.

Now Movsar Barayev hoped to break this stalemate with his bold strike at the
heart of Moscow. His raid had been planned well in advance. In the weeks preceding the attack, the Chechen terrorists quietly slipped into Moscow by train and
car. Their explosives, 120 to 150 kilograms in total, were smuggled into the city on
a passenger bus.

Russian intelligence officials would later determine that some of the weapons
and explosives the group used had been hidden in a nightclub near the theatre by
Chechens involved in a recent renovation of that building.

Barayev's deputy, Abu Bakar, would later tell a Russian TV crew allowed into
the theater shortly after the siege started that some of the terrorists attended *Nord-
Ost* on several occasions to determine the layout of the Dubrovka and finalize
their plans. "We chose the theatre because it is in the centre of the city and there
were a lot of people there," Bakar admitted.

The size of the theatre and the number of terrorists made for a serious problem
for any potential rescue bid by special operations forces. The large cavernous auditorium provided many areas where the Chechens could hide and explosives
had been placed at key locations. To make matters worse, the Chechen women,
dubbed "the Black Widows" by the Russian news media, were wearing belts stuffed
full of explosives. Each of the females grasped a detonator in her hands to ignite a
one-kilogram bomb attached to her waist or chest.

It wouldn't be the first time that a special operations unit faced a situation in which several hundred hostages were being held in a sprawling complex. In December 1996, a group of guerrillas belonging to the Tupac Amaru Revolutionary Movement, or MRTA, had captured much of Peru's ruling elite in a daring raid on the mansion of Japan's ambassador in Lima.

More than a dozen MRTA guerrillas initially took almost 400 diplomats and their families captive but released many as a sign of good faith. Two months later, they were still holding 200 hostages in the large residence as negotiations dragged on.

But luckily for Peruvian special forces, the terrorists became extremely lax in their security. At times, the MRTA guerrillas could hear the sounds of digging underneath the ambassador's residence. Although they rightly surmised that Peruvian commandos were tunneling below them, in their complacency they did nothing to deter the underground operation or even make demands that it stop immediately.

At one point, the terrorists' leader, Rojas, even suggested to Anthony Vincent (Canada's ambassador to Lima who had been acting as a liaison between the Peruvian government and the MRTA) that negotiations might move more quickly if he could arrange a VIP visit to the besieged mansion. The VIP in question was German supermodel Claudia Schiffer, who was visiting Lima at the time. The starstruck terrorists were hoping the blonde beauty would drop in to meet them. Rojas went so far as to suggest that such a visit might go a long way to defusing the hostage situation.

When Peruvian special forces began their assault on April 22, 1997, they caught the terrorists totally off guard as they played soccer in one of the expansive rooms in the ambassador's residence. Casualties to the Peruvian assault force were light considering the danger the soldiers faced: along with two commandos who were killed by a terrorist's grenade, half a dozen were injured. One hostage died from a heart attack but the others were unharmed.

Back in Moscow, Barayev's Chechens were deadly serious at all times. There would be no complacency here, no soccer games to pass the time.

The 25-year-old Barayev was a seasoned Chechen fighter who, when not fighting the Russians, had helped run his uncle Arbi's kidnapping and extortion ring. Arbi Barayev, who had operated a 300-member private army, earned a reputation as one of Chechnya's most ruthless warlords. In 1998, he allegedly ordered the abduction and murder of three Britons and a New Zealander who were working in the country. The men were decapitated and their bodies dumped on the side of a road. Barayev was killed in June 2001 after Russian commandos conducted an eight-day mission aimed at hunting him down.

Barayev's nephew had his own reputation for savagery. In Russia, Movsar Barayev was infamous for sending the relatives of those he kidnapped videotapes of their loved ones being tortured. One video which aired on Russian TV showed a smiling Barayev with a knife at the neck of an unidentified woman. Another video, seized by Russian forces during a raid, reportedly featured Barayev beheading a Chechen woman accused of being a traitor.

Movsar Barayev was also behind several attacks that had taken their toll on Russian forces. In June 2000, a female suicide bomber associated with his gang drove through the gates of a military base in a truck packed with explosives. Seventeen soldiers were killed when she detonated her bomb. Barayev was also believed to be behind another truck bombing, this time in December 2001, which killed 20 Russian soldiers and wounded 17.

The Russians had tried to kill the young Chechen on several occasions, announcing erroneously at least twice in 2001 that he had been gunned down by security forces.

Initially, the Western news media portrayed the Dubrovka theatre siege as another chapter in Chechnya's fight for independence. But for Russian security forces, the hostage-taking was yet another example of the extremes to which Muslim terrorists would go to further their cause.

Intelligence officers had watched with concern over the years as the Chechen rebel movement became increasingly radicalized and connected to the larger network of Islamic extremists. Some guerrilla leaders were now spouting slogans worthy of Osama bin Laden. Mujahedeen from Bosnia, Afghanistan, Iran and Saudi Arabia had also joined up with their Chechen comrades to fight the Russians in what they saw as a Holy War.

Russian intelligence services maintained that there were almost 1,000 al-Qaeda-associated fighters operating in Chechnya. Some of those had gone back and forth between the republic and camps in Afghanistan where they helped train al-Qaeda and Taliban forces.

Some analysts, however, argued that the Chechens still had as their main goal the establishment of an independent country and that any extremist Islamic influence in their movement was minor. Others noted that it was the Russian government's disastrous handling of the 1994 Chechen war, allowing widespread brutality that targeted mainly civilians, which had succeeded in radicalizing the Chechens and in creating a power vacuum which al-Qaeda and related groups took advantage of.

Whatever the reasons, Chechen fighters had indeed turned up in global hotspots to support their Islamic brothers. In the Bosnian war, a large number of Chechens had served with Muslim forces. Chechen Taliban had fought at the uprising at Qala-i-Jangi in Afghanistan, and U.S. special operations forces with Task Force K-BAR and Task Force 11 had killed dozens of Chechen al-Qaeda during their various missions. At the U.S. detention centre at Guantanamo Bay, Cuba, more than two dozen Chechens, taken captive by special operations forces during fighting in Afghanistan, were being held.

The Moscow theatre siege, the Kremlin believed, was no different than al-Qaeda attacks on U.S. targets around the world.

As negotiators tried to reason with Barayev and convince him to release the women and children being held hostage, Russian commandos were already moving into position. The lead unit would be Alpha Group, the country's premier counter-terrorism force created in the 1970s by the KGB to respond to incidents inside the Soviet Union and the Warsaw Pact.

Alpha was comprised of about 250 operators; a core group of about 200 assault troops with backup from specialists who handled sniping, hostage negotiations and underwater missions. The unit had a relatively good track record of operations. It had taken part in successful rescue missions of Soviet POWs in Afghanistan and had emerged relatively unscathed from that war.

With the breakup of the Soviet Union, Alpha found itself dealing with criminal gangs who had been grabbing hostages for profit rather than political motives. It had also gone up against Chechen rebels on several occasions with mixed results.

In June 1995, as many as 100 Chechen guerrillas seized a hospital and surrounding office buildings in the city of Budennovsk near the Chechen border. More than 1,000 hostages had been grabbed. The guerrilla group's demands were similar to those made by Barayev – they wanted the Russian army to immediately pull out of Chechnya.

Russian police and soldiers, including Alpha, stormed the hospital and more than 130 people died in the fighting, although the exact figure of hostages killed is still in dispute because of government secrecy surrounding the incident. Among the dead were 36 police and soldiers, including Alpha members. The Chechens had managed to escape into the mountains with some hostages who they later set free.

Other Alpha missions against the Chechens went more smoothly. In July 2001, a lone gunman seized a bus, demanding the release of five of his fellow Chechen terrorists who were being held in prison. Alpha stormed the vehicle, killing the man and freeing the hostages who escaped injury.

Joining Alpha operators at the Dubrovka theatre were members of the Vympel group, another SOF team created by the KGB intelligence agency in the late 1970s. Vympel troops had originally been trained to carry out assassinations and sabotage in case the Cold War ever got hot, but in the 1990s the unit turned its attention to fighting the organized crime groups that had sprung up with the demise of Communism.

It was apparent almost from the beginning of the siege that the Alpha and Vympel operators would need all their specialized skills to handle Movsar Barayev and his terrorists. The Chechens weren't shy about telling the hostages they fully expected to die in the Dubrovka.

The Black Widows, in particular, seemed resigned to their fate. The women were the widows of Chechen fighters and had taken part in the theatre attack to avenge their husbands. "We're here to die," one of the females told hostage Olga Chernyak, a journalist with the Russian Interfax news agency.

Unknown to those in the Dubrovka, Russian security forces had infiltrated into the building just hours after the siege began. In the basement, officers from the Federal Security Service, or FSB, the successor to the KGB, had set up listening devices to monitor the activities in the auditorium. They could hear the terrorists talking among themselves and by carefully monitoring the movements above them were able to pinpoint most of the locations of the gunmen.

The FSB determined that the Black Widows had strategically spread themselves out around the auditorium. Several were sitting among the children who were grouped in seats underneath the balcony. If they detonated their bombs, there was a good possibility that the upper structure would collapse and crush the children. Several other women were positioned near the larger 45- to -50-kilogram bombs.

Barayev had made some concessions in the early morning hours of Thursday by releasing at least 40 people, including some women and children. But later in the day, the situation deteriorated. Although the terrorists had allowed the hostages to use their cell phones to contact family members and in some cases give interviews to the news media, one of the Chechens shot and killed 26-year-old Olga Romanova after she refused his order to immediately stop talking on her phone.

Tension continued to increase when two women, who had been allowed to go to the bathroom, slipped out a window and escaped the Dubrovka without injury, even as the Chechens threw grenades at them as they fled.

From then on, the Chechens decreed that the orchestra pit would be used as a toilet. In no time, the stench from hundreds of people using the makeshift lavatory was overpowering. Bright lights in the auditorium were kept on night and day and few people got any sleep. There was little water and the only food given to the hostages was some chocolate. For the most part, the people were confined to their seats.

Hostage Maria Shkolnikova, who spoke to Moscow Radio on her cell phone, described how explosives had been placed throughout the theatre. "People are close to a nervous breakdown," she warned.

The Qatar-based al-Jazeera TV network identified the hostage-takers as the Martyrs of the 29th Division and aired a videotape of Barayev declaring: "I swear by God we are more keen on dying than you are keen on living. Each one of us is willing to sacrifice himself for the sake of God and the independence of Chechnya."

Preparations for a rescue raid, being considered from day one, were accelerated. Intelligence that was being gathered for the mission received a major boost thanks to an FSB agent who happened to be attending the *Nord-Ost* performance the night of the attack. Now being held hostage, he used his cell phone to discreetly relay information about how the Chechens had set up their bombs. Although he wasn't 100 per cent sure, he believed the two largest explosive devices had likely not yet been armed. The FSB agent estimated it would take the Black Widows at least 60 seconds to activate each bomb; a process that involved wiring up a car battery to the detonator and then pressing the ignition button.

An Alpha team was able to confirm the agent's information when its members slid a miniature TV camera through a hole they had quietly drilled into a wall of the auditorium.

Across the city at another theatre, other Alpha operators were practicing their assault drills and meticulously running through the plan of how they would conduct their raid. Vympel operators played the role of the Chechens in the training scenarios.

It was determined that because of the size of the theatre, an assault force of 200 commandos would be needed. That group would be broken down further into teams of four operators; each team was given a specific terrorist to target and eliminate in the opening seconds of the raid.

The main problem for the takedown was the suicide bombers, namely the Black Widows, who never seemed to let go of the detonator buttons rigged to the explosive belts they wore. Especially critical was the need to prevent any of the terrorists from hooking up and detonating the larger bombs.

The Russians had taken a look at the construction blueprints for the theatre and determined that the structure could not withstand the power of such a blast. There was a good chance the roof would collapse, crushing many of the hostages.

At the same time, the Chechens were savvy enough that each time they received a visit from a Red Cross worker or negotiator they relocated some of the smaller explosive charges they had rigged in the auditorium. That way, any rescue force would not know for sure where all the bombs had been hidden.

Barayev had already warned what would happen if there was any rescue raid. In a statement attributed to him and posted on a pro-Chechen website he wrote: "No one will get out of here alive and they'll die with us if there's any attempt to storm the building."

To deal with that potentially deadly outcome, the raid planners decided to use a gas to incapacitate the terrorists in the hope they would be knocked unconscious before they could blow themselves up.

The Russians decided to employ an aerosol containing Fentanyl, an opiate commonly used as an anesthetic. Fentanyl, which is usually administered by injection, could make a person unconscious in less than a minute. But in high enough doses it could also be fatal and proper use of the drug had to be dictated by a person's size, weight and overall health. The Russians were breaking new ground in that they would try to administer the narcotic in a spray to a large and diverse group of people.

By Friday, the Chechens were on edge and growing increasingly angry that there appeared to be no movement to meet their demands. The terrorists had been taking drugs to stay awake and were growing more hostile toward the hostages. At one point, there was a loud noise in the theatre, startling the Chechens who thought a rescue mission might be underway. The Black Widows reacted immediately by positioning themselves around the auditorium at six-metre intervals, their fingers ready on their detonator switches. They only relaxed after a few minutes when it became clear that an assault was not in progress.

Barayev, angry over the lack of progress in the negotiations, announced he would start killing some of his captives on Saturday if he didn't see evidence that Russian troops were getting ready to pull out of Chechnya. In an attempt to delay the executions and give Alpha more time to work out its rescue plan, Russian negotiators told Barayev they had arranged for him to talk with Viktor Kazantsev, the senior Russian army general in Chechnya. The meeting had been set for 10 a.m. on Saturday.

Barayev accepted the invitation, although somewhat leery of the gesture. He released another 19 hostages and announced that the executions would be postponed until noon on Saturday.

Rescue planners huddled among themselves to take stock of the situation. The primary concern among intelligence officers who had been monitoring the

Chechens from the theatre basement was that the terrorists seemed confused about what to do next. They didn't appear to have any real plan beyond seizing the theatre and taking hundreds of hostages in the hope the Russians would meet their demands. Strangely, there was little talk among the Chechens about escaping after their mission was finished.

Also disturbing were the comments that Barayev had made to some of the negotiators. He had become almost impossible to talk to and seemed determined to live up to his word to kill hostages on Saturday.

"Tomorrow, I will meet Allah," Barayev told negotiator Yeugeny Primakov. "I am afraid of nothing."

In the early hours of Saturday morning, the Black Widows began to play a tape recording of Muslim prayers. Some hostages were so convinced they would die in the next few hours that they wrote messages in pen on their bodies for their loved ones to find.

Outside, snipers waited in their positions on rooftops covering the entrances to the Dubrovka while troops patrolled the streets in armored personnel carriers.

What took place early Saturday to cause Russian officials to launch the rescue raid is still in dispute. Some hostages claim the situation was calm in the theatre just before the attack. Others say that the terrorists had shot two of their captives in the early morning hours. Hostage Olga Chernyak, the Interfax journalist, said that a boy, overwrought from lack of sleep, picked up a bottle and threw it at one of the Chechens. As he ran away, crying "Mommy, I don't know what to do," some of the Chechens opened fire on him. Their bullets missed the boy but hit two hostages who were in their seats.

Whatever the real story, shortly after 5 a.m. Alpha and Vympel radios came alive as the "go signal" was transmitted to all operators. The gas cylinders' valves were turned on and the aerosol containing the Fentanyl was pumped into the theatre. Estimates of how long it took for the powerful drug to fill the auditorium vary from 15 minutes to half an hour.

Some of the hostages would later say they felt a tingling in their noses and then fell unconscious. Others didn't remember a thing and only regained consciousness once in hospital. A few recall seeing the gas, which they described as a grey or green cloud of mist.

Andrei Naumov, a 17-year-old dancer with the *Nord-Ost* troupe, slowly felt his body going numb. He was alert for 10 to 15 seconds and then slumped down to the floor. As he was slipping into unconsciousness, the young man thought to himself that a rescue raid was about to be launched.

Someone in the auditorium shouted, "Gas! Gas!" and several terrorists put on gas masks. Others collapsed almost immediately, their fingers slipping harmlessly off the detonators rigged to their explosive belts.

Even as the gas was being pumped into the auditorium, three Alpha assault teams, including one that came up from the basement, entered the theatre. The red beams from the laser-targeting systems on their machineguns criss-crossed the auditorium in search of terrorists.

Some of the Chechens had made it to the auditorium's stage and were firing their assault rifles at the approaching commandos. The terrorists collapsed like rag dolls as their bodies were hit by a hail of gunfire. In the lobby, several Chechens managed to throw a couple of hand grenades before an Alpha team shot them dead.

In the hall next to the theatre entrance, a lone Black Widow, a pistol in her hand, a grenade in the other, and a belt of explosives around her waist, managed to get off just one shot at the commandos before they riddled her body with machinegun fire.

Upstairs in the theatre offices, another group of Chechens had barricaded themselves as the assault teams approached. An Alpha member booted open the door and several grenades were tossed in, killing all inside.

Some of the hostages managed to break free and escape from the auditorium. Just after the shooting started, five women, covering their faces with rags as protection against the Fentanyl gas, ran from the theatre.

Some of the Alpha operators took off their gas masks after finding the devices severely restricted their ability to fight. When the unmasked commandos felt they were being overcome by the effects of the Fentanyl, they would stick their fingers down their throats, vomit, and continue shooting. This tactic didn't work all the time since a few of the special forces operators passed out after inhaling large doses of Fentanyl. Nine of the men were later treated for exposure to the gas.

As the firefight subsided, members of the Alpha team moved from one terrorist's body to another, pumping two bullets into each of their heads. Some of the women were reportedly still alive but unconscious from the gas when they were executed. A team member would later describe the tactic as a "control shot."

Amazingly, none of the Black Widows were able to detonate the explosive belts attached to their bodies. Every single terrorist had been accounted for and killed. The Russian news agency Pravda would later report that when Barayev's body was found he was clutching a bottle of cognac, the less than subtle insinuation being that he was not exactly a devout Muslim.

By 8 a.m., Alpha declared that it had control of the theatre and explosive ordnance specialists started removing the numerous booby-traps the Chechens had rigged in the building. They worked around the dead terrorists, some still slumped in their theatre seats, their heads titled back and their mouths grotesquely hanging open.

Outside the scene was chaos, as hostages, some dazed, others still unconscious from the Fentanyl, were dragged to ambulances and buses. Initially, Russian officials said that all the hostages who had died were killed by the Chechens. A government spokesman informed the news media that terrorists had executed 67 of the hostages as commandos stormed the building.

A few hours later, the death toll had jumped to about 100, but as that figure climbed even higher, word started leaking out that those who died had succumbed to the effects of a mystery gas and not to terrorist bullets. Hundreds more were in hospital being treated for exposure to the narcotic.

One Canadian military analysis later concluded that it would have been extremely difficult to figure out how much gas would quickly incapacitate the healthy adult terrorists without risking an overdose to pregnant women and children being held hostage. "To have executed this operation in the face of dozens of terrorists equipped with suicide bombs, holding hundreds of hostages in a cavernous structure, without a single civilian casualties (sic) would have been nothing short of miraculous," it noted.

Indeed, there would be no miracles.

As the Russian government slowly acknowledged that the hostage deaths were the result of the gas, its officials went on the defensive. Health Minister Yuri Shevchenko insisted that under normal circumstances the gas would not have been fatal. But he said that the conditions the hostages lived in – almost 58 hours in captivity without food or water and under incredible stress – had weakened some of them enough that exposure to the gas was fatal.

The Kremlin was also criticized for refusing, at least initially, to tell doctors treating the hostages what type of gas they had been exposed to.

Not all, however, thought the actions of the Alpha team were unreasonable. One of the rescued hostages, Nikholai Zhizhen, said that because of the large number of explosives scattered throughout the Dubrovka, the assault teams had no choice but to use the gas.

The reason it was used in the first place, according to some Alpha operators, was because of the concern that a group of fifth columnists had pretended to be hostages. This team of up to six Chechens, selected because of their Slavic looks and ability to blend in with Russians, had purchased tickets for *Nord-Ost* shortly before the Wednesday attack and were secretly situated among the hostages, the Russian special operations soldiers claimed. Their mission was to keep an eye on the captives and warn Barayev about any escape attempts. They were also to act as a backup for the Black Widows in case the women were somehow prevented from detonating their explosives.

The concern among the Alpha team was that it would have been impossible to locate these individuals in the crowd once the raid began – and at that point it might have been too late to prevent any damage they might do. So the gas was needed to incapacitate as many people as possible in the auditorium before the assault started. It is not known for certain what happened to the fifth columnists, or if they actually existed.

A special operations soldier told the Russian newspaper *Moskovsky Komsomolets* that the gas had been developed by the KGB and had been used before by the Russian Army, although he did not provide details. The operator also noted that the U.S. Delta Force and German GSG-9 counter-terrorism team had similar substances in their weapons inventories.

As the bodies were being pulled out of the Dubrovka, Russian President Vladmir Putin went on TV to address the nation and apologize that the raid wasn't able to save all the hostages. He also thanked members of Alpha and Vympel as well as the other security forces who took part in the operation. "They have proved that terrorists will not force Russia to its knees," said Putin, a hard-liner when it came to Chechnya.

The death toll would eventually climb to 128. Five of those included hostages who had died from bullet wounds in the auditorium, but the circumstances surrounding their deaths were cloaked in secrecy and have never been adequately explained. One Russian report stated that the five were executed by the Chechens.

There has also been much speculation about whether the death toll was higher than 128. Rumors still abound in Moscow that anywhere from 200 to 300 hostages were actually killed from the gas used during the raid.

Human rights activists and families of the dead hostages protested how the raid had unfolded. The activists alleged that the Chechens had been summarily executed, while relatives of the dead hostages called the assault a poorly planned debacle.

Criticism, however, among other nations was muted. The U.S. government, once a vocal opponent of Russia's war against the separatists in Chechnya, said it understood and sympathized with the Putin administration's decision to undertake the raid. "The United States condemns terrorist attacks wherever they occur, and no political grievance justifies the taking of hostages and killing of innocent people," said State Department spokesman Richard Boucher.

The response was not surprising. Russia had become a valued ally in America's war against al-Qaeda, providing intelligence as well as remaining silent when U.S. special operations forces used former Soviet republics as staging areas for their missions into Afghanistan. In the weeks following the September 11 attacks, the Kremlin had also authorized Russian officers to provide the Americans with maps and other details on fortifications and caves they had encountered during their war in Afghanistan. It was useful intelligence that could be passed on to U.S. special operations forces in their hunt for the enemy.

Bush administration officials were also becoming more aware of the potential al-Qaeda-Chechen connection. Several months before the raid, an American chargé d'affaires in the former Soviet republic of Georgia told journalists that dozens of al-Qaeda-linked terrorists were known to be operating in a region in the country known as the Pankisi Gorge. The Arabs had been living in camps under the control of Chechen rebel groups. French intelligence had also discovered links between Islamic rebels in Chechnya and Zacarias Moussaoui, who was alleged to have been in on the September 11 plot. A Moroccan who held a French passport, Moussaoui had been arrested in the U.S. where he was taking flight training and charged with conspiracy in the attacks.

Western counter-terrorism and SOF units would study the Russian assault in detail, noting what had worked and what had failed. But those in the U.S., Britain and Australia wouldn't have much time to linger over the lessons learned in Moscow. They had been given a new target in the war on terror: Iraq.

OPPOSITE PAGE: Blowing sand and scorching temperatures were common challenges for all coalition SOF in Iraq, including these Australian SAS troops. (COURTESY ADF)

Turning Point: Operation Iraqi Freedom

7 The Australian SAS trooper hoisted a Javelin anti-tank missile and took aim at one of the two Iraqi trucks speeding towards him.

As the Iraqis got closer, they opened fire with RPGs and machine-guns. The 20 Iraqi commandos, travelling in two SUVs, were determined that the missile installation they were guarding wouldn't fall into the hands of coalition forces.

Standing in the turret of his long-range patrol vehicle, the SAS soldier looked into the Javelin's sighting system and manipulated its computer cursor until it was precisely on top of the image of one of the approaching SUVs. He then pressed the firing button, sending a missile flying downrange at more than 90 metres a second. As it cleared the launch tube, the missile's infrared seeker took over, homing in on the truck's engine heat. When the eight-kilogram explosive warhead slammed into the SUV, it lifted the truck momentarily into the air before the vehicle burst into flames.

As the first vehicle lay burning, 36-year-old Trooper X – the name given to the Australian in later dispatches of the day's event – swung the patrol vehicle's .50 calibre machinegun around and opened fire on another Iraqi position. Bullets whizzed by as enemy commandos directed their gunfire towards him. Trooper X quickly prepared another Javelin to take out the second Iraqi vehicle. Again the missile did its job, turning the truck into a twisted hulk of smoking metal.

Seeing that other Iraqi soldiers were now preparing a mortar for firing, Trooper X grabbed a sniper rifle, put the crosshairs on the weapon, and squeezed the trigger. His first shot hit the mortar tube, disabling the device and sending Iraqi soldiers diving for cover. By now, some Iraqis were surrendering to the advancing SAS soldiers, but others determinedly held out, continuing to shoot their AK-47s at the Australians. After Trooper X placed several well-aimed shots at the Iraqi holdouts, they finally surrendered, realizing the SAS man had them in his deadly sites.

He probably didn't think about it at the time, but Trooper X was helping make special forces history. Although the war in Iraq would last less than two months, it would be a turning point for SOF units. Operation Iraqi Freedom, the code-name for the war of 2003, would witness the largest use of such soldiers since the Second World War.

For the Iraqi Freedom mission, an estimated 10,000 special operations troops were put into the field. There were more than 250 Navy SEALs, the unit's largest single deployment since Vietnam. Another 250 naval special warfare personnel, mainly crews who manned the SEALs' ultra-fast Mark V craft and rigid inflatable boats, were on hand for support. The British provided an estimated 300 SAS and an unknown number of SBS, while the Australians committed 150 of its SAS operators. The Australian SAS was also backed up by the country's 4th Battalion, Royal Australian Regiment (Commando), which acted as a quick-reaction force for the operators. The Polish GROM, (Grupa Reagowania Operacyjno Mobilnego or Operational Mobile Response Group), which had worked with the SEALs in Haiti in 1994, contributed 56 men. Hundreds of Green Berets and U.S. Air Force combat controllers were also on the ground, as well as squadrons from Delta Force and personnel from the CIA's special operations branch and other intelligence spooks. Rounding off the contribution were other units such as the 160th Special Operations Aviation Regiment and the U.S. Army Rangers.

"They are more extensive in this campaign than any I have seen," U.S. Army Major General Stanley McChrystal, the vice director of operations of the Joint Chiefs of Staff, said as the conflict was being waged. "Probably, as a percentage of effort, they are unprecedented for a war that also has a conventional part to it."

Taking out Saddam Hussein's regime in Iraq factored into the American war on terrorism even as it was in the planning stage in the days following the attacks on the World Trade Center and Pentagon. President George W. Bush told his advisors he was convinced that Saddam was somehow involved in the attacks. Defense Secretary Donald Rumsfeld and his deputy, Paul Wolfowitz, also strenuously pushed for including Saddam as a target in the global war on terrorism.

For the Americans, the possibility that Saddam would provide weapons of mass destruction technology to U.S. enemies – in particular al-Qaeda – was enough to seal the Iraqi leader's fate. Although it would be a highly controversial issue long after the war ended, American officials claimed that Iraq had provided training to al-Qaeda operatives in how to develop chemical munitions. Saddam had also continued to defy the United Nations, according to Bush, by secretly continuing a program to develop WMDs. Leaders from other countries, however, were at odds with the president on the need for military action. They continued to argue that a program of UN inspections of suspected Iraqi weapon sites should be allowed to continue.

Because of the controversy over WMDs, several countries that had been quick to provide support in the war in Afghanistan were less inclined to help in Iraq. This time around, the U.S. commandos would not be joined by comrades from Canada, Germany, New Zealand, Turkey or Norway – some because of the WMD controversy, others for domestic or military reasons.

Whatever the truth about WMDs, by late 2002 it was clear to the U.S. that Saddam's days were numbered. Fresh from the conflict in Afghanistan, CENTCOM General Tommy Franks was given the job of planning the Iraq invasion. With the Afghan experience behind him, and with considerable influence from the pro-special operations Rumsfeld, Franks designed his battle plan from the outset with such units in mind. Representatives from SOCOM, including Brigadier General Gary Harrell, along with CIA officers, were situated in Franks' headquarters to improve communication between conventional and special operations units as well as to co-ordinate missions.

The main objectives of the Operation Iraqi Freedom campaign were to dismantle Saddam's regime and the country's WMD program. To meet these, Franks' plan was to approach the country on several fronts at the same time, using SOF wherever appropriate.

In the west, the general planned to send commandos in to hit suspected Iraqi Scud missile launch sites. In this respect, U.S. concerns hadn't changed much from the 1991 war. The Israeli city of Tel Aviv was within firing distance of the Scud bases in the western desert, and Franks wanted to keep Israel out of the battle. Doing that meant reassuring the Jewish state that the U.S. would take care of any Scuds. The Pentagon had overruled an Israeli plan to put their own special operations forces into Iraq, but as a compromise it had established a command center in Israel where the country's leaders could monitor the missile-hunting missions. Franks' decision to use special forces in the western desert also negated the need to commit at least a division-sized conventional unit there.

A second front would be created in the north by having the U.S. Army's 4th Infantry Division invade from Turkey and seize the Kirkuk oil fields. But Turkish political leaders scuttled that plan when they refused to allow large formations of American troops into their country. The flexibility offered by special operations forces soon became apparent as Franks dealt with this setback by using SOF, who had already slipped into Iraq, to lead Kurdish Peshmerga fighters in opening up a northern front. Those forces were backed by troops from the 173rd Airborne Brigade who dropped into the north in what was described as the largest American combat parachute assault since the Second World War.

In the south, forces would make their way toward the capital city of Baghdad. The key job of securing the coastal port city of Umm Qasr would fall to SEALs, GROM, Royal Marines from Britain, and the U.S. Marines.

Finally, a highly secretive unit dubbed Task Force 20, made up of Delta and SEAL operators, as well as U.S. Army Rangers, would be charged with finding WMDs, capturing or eliminating the Iraqi leadership and dealing with other "high-value" targets.

Preparations for the war had been going on in earnest since November 2002. Even before the conflict began, British and U.S. fighter aircraft had been pounding Iraqi air defences. (Publicly, Pentagon officials said they were retaliating against Iraqi attempts to shoot down the aircraft, but after the war they would acknowledge the missions were specifically aimed at destroying air defence systems in preparation for Operation Iraqi Freedom.) Special operations forces started slipping across the Iraqi border for a variety of missions. Some conducted reconnaissance on potential landing areas, while others gathered information on likely targets. CIA teams, as well as agents from the British intelligence service MI6, started making contact with potential allies inside the country.

As well, SOCOM quietly approached retired operators who had worked on Operation Provide Comfort in 1991. That mission had been to deliver humanitarian aid to the Kurds whose ill-fated rebellion against Saddam's regime had been savagely put down by Iraqi troops. The retired special operations soldiers knew the terrain and the Kurdish leaders and seemed the obvious choice to help launch combat operations in the north. Almost 90 men volunteered to go back into service and cross into Iraq, a testament to their patriotism – and likely to their sense of adventure.

In some cases, the preparations for war weren't all that secret. On February 13, 2003, a convoy of cars carrying news reporters followed a caravan of special operations SUVs as they were leaving Turkey and heading into Iraq. Some of the media cars roared up beside the SUVs, filming the operators inside – most likely

Green Berets and Air Force combat controllers – who tried to hide their faces.

In early March, the Australian and British SAS along with Delta Force moved into the Azraq airbase in Jordan. By the time the war began, there were already an estimated 150 to 200 operators from coalition military and intelligence agencies inside Iraq.

Some of the insertions into the desolate desert proceeded smoothly while others were more hairy. Two days before the March 20 launch of the invasion, the first Green Beret A-team headed into the southern Iraqi desert in an insertion that almost ended in disaster. On the night of March 18, three MH-53 Pave Low choppers carrying the A-team and their Toyota pickup trucks flew over the Iraqi border, skimming low across the rocky terrain at high speed. The first chopper came into the landing zone, kicking up massive dust clouds. When the other Pave Lows moved in for a landing, their pilots, wearing night-vision goggles, could barely see through the swirling clouds. One of the helicopters clipped the top of a rock tower, sending the bird out of control and crashing onto its side.

Overhead a Predator UAV circled, beaming a live video picture of the crash scene to officers at a base in Kuwait. There was chaos for the first few moments as radios crackled with "Chopper down, chopper down," and crews tried to determine the extent of damage and whether there were any injuries. Amazingly, the crew and special operations soldiers aboard escaped the crash with no major injuries, but the aircraft and a truck it was carrying were written off. Like true professionals, the men unloaded the other vehicles and moved out toward their target, the city of Nasiriyah.

By the night of Thursday, March 20, the war was underway. Navy SEALs launched their first major mission of the war by seizing a pumping station and refinery at Al Faw in southeastern Iraq. The refinery was a critical target. War planners were worried that Saddam Hussein might order the complex destroyed, crippling the chances that any new regime might have of putting Iraq back on its economic feet. There was also the possibility that the Iraqi leader might revert to the environmental warfare he instituted during Desert Storm in 1991 when he ordered refinery oil taps to be opened and flooded parts of the Persian Gulf with crude. At the helm of the Al Faw mission was Navy Captain Bob Harward, the SEAL who had led Task Force K-BAR in Afghanistan.

The plan called for the SEALs to seize the pumping station and hold it until Royal Marines and units from the U.S. Marine Corps reached their position. The operation had been in the works for weeks. U.S. Navy P-3 Orion maritime patrol aircraft and Predator UAVs had already flown over the facility, taking video and photographic images so the raiding force could map out its attack. The SEALs would have a delicate job of taking out the Iraqi bunkers around the complex while avoiding touching off large explosions that could set the pumping station ablaze.

Five Pave Low helicopters carrying the 100-member SEAL strike force descended on the Al Faw complex shortly before midnight on March 20. Each helicopter hovered over one of the five key locations in the sprawling complex and the commandos fast-roped down onto their targets.

AC-130 gunships provided air cover, hitting Iraqi bunkers and troop formations on the ground while RAF fighters used smart bombs to take out anti-aircraft gun installations. An EC-130 electronic warfare aircraft flew overhead jamming Iraqi radio communications while a P-3 Orion provided real-time video to commanders watching the raid unfold back at their headquarters in Kuwait.

The SEAL strike force conducted its mission without casualties and seized the key points of the installation, such as pumping machinery and controls, within the first 15 minutes of the assault. By dawn, the oil facility was secure. Although the Navy declined to release the numbers of Iraqis killed in the raid, some reports put it between 50 and 100 enemy dead. Most Iraqis, however, surrendered or fled when the shooting started.

At the same time that the pumping station was being seized, SEALs and commandos from the Polish SOF unit GROM were using high-speed boats to get on board two Iraqi offshore oil terminals 20 kilometres off the coast in the Persian Gulf. The terminals, Kaabot and Mabot, were massive structures, each about a kilometre and a half long. SH-60 Seahawk helicopters from aircraft carriers USS *Constellation* and USS *Abraham Lincoln* circled the platforms in case the SEALs needed support. But the Iraqi soldiers on the oil installations surrendered without firing a shot. One of the Seahawks used its infrared sensor to transmit live images of the 15-minute takedown to a SEAL command post on the cruiser *Valley Forge*.

Later, the SEALs started clearing the waterway leading to Umm Qasr, Iraq's only deep-water port. Although the battle for the city would go on for days, clearance operations got underway almost immediately since the coalition plan called for relief supplies to be moved into the Umm Qasr and distributed throughout the country.

The SEALs were also patrolling the waterway for escaping senior officials of the Iraqi regime and to thwart any attempts by Saddam's forces to mine the approaches to Umm Qasr. One SEAL patrol seized an Iraqi tug carrying more than 80 seamines while a Navy P-3 Orion detected several Iraqi navy patrol boats that were possibly laying mines. The air crew relayed their information to an AC-130 gunship which destroyed the vessels with 105 mm howitzer fire.

Also problematic for the SEALs were the dozens of rusting and half-capsized ships along the waterway, some left over from air strikes during Desert Storm, others victims of Iraq's faltering economy. Still other ships, not damaged, appeared to have been abandoned. But all of them had to be searched since they could be used by Iraqi forces to ambush coalition supply ships travelling along the waterway. Using their high-speed Mark V boats and rigid inflatables, the SEALs, along with GROM commandos, checked out the vessels one by one. At times, Iraqi militia, hidden along the shore, opened fire on the teams. Some of the most tense missions took place at night as masked GROM operators, covered by SEALs manning .50 calibre machineguns on board the Mark V boats, would carefully board the abandoned ships and conduct cabin-by-cabin searches for enemy forces.

Meanwhile, GROM hooked up with British Royal Marines to conduct hit-and-run raids in Umm Qasr, eventually playing a key role in the fall of the city. The unit would later get into trouble back home when photos were shown in which some operators posed while handing over Iraqi prisoners to the Americans. They were also photographed defacing a portrait of Saddam Hussein and posing under an American flag.

The Polish public loved it, but Defence Minister Jerzy Szmajdzinski was livid. While admonishing the commandos for supposedly violating security, his main concern appeared to be that his government had been caught in an embarrassing lie about the Polish military's true role in Iraq. The government had originally denied that GROM was involved in combat. But after the photos appeared in newspapers and amid allegations that it had misled the public, the government reluctantly had to acknowledge the obvious.

Szmajdzinski continued to refuse to give details about the missions, other than the fact that GROM was operating in the "coastal region" of Iraq. "These photos shouldn't have happened," the Defence Minister said.

The SOF campaign highlighted the wide range of activities at which such units excelled. Air Force combat controllers successfully targeted Iraqi tanks and armored columns. Inside Baghdad, members of Delta Force scurried along the city's dank, putrid sewer system, attaching listening devices to underground telephone lines or cutting through fibre optic cables to prevent Iraqi leaders from communicating with each other.

Some of the most important covert work began in the days leading up to and shortly after the war started. Operators from MI6, the CIA and the British SAS made deals with some Iraqi generals, after presenting them with limited options: keep your troops in the barracks and you will receive ample financial reward; resist, and the Iraqi forces will be destroyed.

At Basra, on the same day the SEALs hit the Al Faw pumping station, a 12-man Green Beret team hooked up with several hundred followers of a Shiite cleric and organized this anti-Saddam group into a partisan force. Arming them with Chinese AK-47s and RPG-7s, the A-Team began a series of hit-and-run raids with the Shiites against Iraqi army targets, including attacks on ammunition and weapons dumps.

One of the biggest SOF operations took place in the opening days of the war in the western desert near Iraq's border with Jordan where a joint force of 100 to 150 British SAS and 100 Australian SAS seized the airfields dubbed H2 and H3. Members of Delta Force, U.S. Army Rangers, paratroopers from the 82nd Airborne and U.S. Air Force combat controllers were also involved in the fighting over a two-day period.

U.S. planners were worried that the bases, located near the desert town of ar-Rutba, 380 kilometres west of Baghdad, could be used to launch Scud attacks against Israel. Satellite photos of the area indicated the Iraqis had moved truck convoys and what appeared to be missile launchers into the bases. Seizing the bases would neutralize the Scud threat and also give the coalition a secure location for Apache attack helicopters and other aircraft to provide close air support for an advance on Baghdad.

The airfields were targeted even before the war started, being hit on the morning of March 14 by two B-1B bombers in an effort to knock out a mobile early-warning radar and air defence center. It is believed the bombers were guided to their targets by Australian and British SAS who had already been inserted by an MH-47 Chinook helicopter to conduct initial reconnaissance on the airfields.

The main British SAS force, made up of more than 100 troops, seized the bases in a highly mobile attack involving dozens of Land Rovers which spread out among the buildings and command and control centres. The SAS quickly took over the control tower and went from hangar to hangar engaging in fierce firefights with Iraqi troops. Other SAS troopers engaged in hand-to-hand combat as they killed Iraqis guarding the missile launchers.

Several hundred Iraqis were taken prisoner and scores were killed. After the initial seizure of the airbases, about 1,000 paratroopers from the 82nd Airborne were brought in as reinforcements. Delta Force, the Australian and British SAS, and coalition troops together beat back at least one determined Iraqi counter-attack.

Although no Scud missiles were found, H2 and H3 soon became a hub for coalition air operations as special operations forces continued pushing across the desert towards Baghdad. Royal Air Force Harrier jets and U.S. Air Force A-10s provided cover for the units while Predator UAVs often preceded the commando columns to scout out targets for attack. The Australians, in particular, mounted aggressive patrols to keep Iraqi garrisons in the region off guard. "They are generally creating havoc and uncertainty behind lines, and are constantly redeploying in their area of operations," Australian army chief Lieutenant General Peter Leahy told journalists as the war unfolded. Part of the havoc was created by hit-and-run raids against command and communications centres, as well as targets of opportunity.

Some of the Australian SAS had been inserted 600 kilometres into Iraq by a U.S. helicopter while another group, travelling in a column of long-range patrol vehicles, stole across the border at night from Jordan. The latter was able to successfully avoid a series of Iraqi guard posts, but after travelling just 30 kilometres, the column ran straight into an enemy patrol. After a brief firefight, the SAS rounded up their Iraqi prisoners while medics treated two injured Iraqi soldiers. Not wanting to be slowed down by prisoners, the SAS stripped the men of their weapons and sent them on their way.

Meanwhile, the 600-kilometre helicopter insertion was even more difficult, being carried out at night in poor weather and at low altitude. During the journey, the chopper crew conducted a tricky mid-air refueling, as well as having to avoid extensive enemy air defence systems. "When our people hit the ground, they were at that time by far and away the closest coalition ground elements to Baghdad and they remained that way for a number of days," said Australia's Special Operations Command Chief of Staff Colonel John Mansell, who declined to identify the SAS location, other than to say it was in the western desert.

Once inside the country, the Australian SAS determined that their mission would best be accomplished by conducting aggressive patrols that would draw out Iraqi troops. While those patrols were underway, other SAS teams conducted surveillance on the main highways in case the Iraqis moved Scuds or larger conventional forces in and out of the region. Almost every day, the SAS fought heavy firefights with the Iraqis, with officers noting that the enemy was clearly seeking out the Australian forces in a coordinated fashion.

On their second night in Iraq, a large number of the SAS gathered for an attack on a well-defended radio relay station. A team put the station under surveillance and then called in air strikes to destroy its transmitter tower. SAS officers say a large number of Iraqis were killed although no official numbers have been released. Knocking out that tower, Australian officers noted, was one of the keys to crippling Iraq's ability to launch Scud missiles, although it would eventually be determined that the country never did possess the banned rockets, at least not in the Australian area of operations.

The attack on the radio relay station didn't go unnoticed and Iraqi commanders promptly counter-attacked with troops and six armored vehicles. An SAS team was involved in a running firefight for a number of hours with that force but the Iraqis retreated when the commandos called in air strikes and destroyed the armored vehicles.

At another location, an SAS patrol was set up to conduct surveillance on Highway 10, a main roadway in the area. The troopers were well enough hidden that they went entirely undetected even though there were a number of enemy moving around the area, as well as nomadic Bedouin. It was during such a reconnaissance that the Australians determined that what appeared to be a truck-stop was actually a key Iraqi installation. Dubbed Kilometre 160, the centre was a main marshalling area for Iraqi forces in the region and at one point was defended by more than 200 soldiers. Over a 48-hour period, the SAS team called in air strikes from its concealed position, eventually obliterating the target. The commandos then assembled a large vehicle strike force to attack Kilometre 160. However, Iraqis who survived the bombing were able to withdraw under the cover of a sandstorm.

It took about a week for the enemy in the Australian area of operations to be neutralized and, for at least the first 96 hours, the intensity of the combat was such that the SAS troopers didn't get any sleep. Iraqi casualties in confrontations during the opening days of the war with the SAS were said to be at least 100.

Besides having to deal with the enemy, the desert also offered up its own challenges for the SAS and other special operations forces units roaming the vast expanse. Temperatures at night dropped to minus 5 degrees Celsius while during the day they exceeded 40 degrees. On one occasion, sandstorms blew constantly for two days, ripping up tents and coating vehicles and equipment with sand that had the consistency of talcum powder. At times, visibility was reduced to 10 metres. In contrast, there were other occasions when it rained so heavily that the SAS found its weapons being clogged by wind-blown mud.

Part of the reason the troopers were able to mount their long-range patrols so far behind enemy lines was a well co-ordinated logistics chain. Tonnes of ammunition, water and food were airdropped to the SAS.

Typical was a drop that contained more than 1,500 patrol rations, more than 1,000 litres of water, two tonnes of JP8 fuel, spare parts, radios and batteries, and ammunition including additional Javelin missiles. The gear was packed onto 10 specially designed containerized delivery pallets attached to parachutes and delivered to the SAS by U.S. Air Force special operations aircraft. To protect the equipment, particularly the communications gear, the pallets used honeycombed card-board to cushion the load as it hit the desert floor.

While the SAS, both British and Australian, controlled western Iraq, in the north end of the country, Green Berets and Air Force combat controllers opened up a new front, setting the stage for the 173rd Airborne Brigade's 1,000-man para-chute drop.

One of the main jobs SOF had in the north was to direct air strikes on Iraqi front-line positions outside Mosul and Kirkuk. The situation was similar to that in Afghanistan, where special operations forces would link up with local indigenous troops. This time, it was the Kurdish Peshmerga who would help them select tar-gets and provide most of the ground troops for attacks.

But there were important differences between the Afghan war and the fighting in northern Iraq. For starters, the Peshmerga were more aggressive than the Af-ghan fighters and seen as more trustworthy by the special operations troops. The Kurds and their Peshmerga fighters had been conducting a brutal guerrilla war against Saddam's regime for decades and the Iraqi leader had responded by using chemical weapons as well as mass executions. The Peshmerga were fighting for the establishment of a separate Kurdish country. Unlike Afghanistan, there would be no side deals allowing enemy fighters to escape, no tip-offs to Iraqi forces that an attack was about to commence.

The other difference was the vast improvement in response times by fighter and bomber aircraft to special operations needs. In Afghanistan, it took an aver-age of 45 minutes between the time operators on the ground called for air strikes and the time bombs started hitting their targets. In the Iraq campaign, procedures had been so streamlined that the average time had been cut down to 11 minutes.

Besides the main cities in the north, one of the key targets for special opera-tions troops were the training camps of the Ansar al-Islam (Followers of Islam) terrorist organization. Ansar had established itself in Iraq several years before, imposing strict Islamic rule on the dozen villages it controlled near the border with Iran. The Bush administration maintained that the group had direct links with al-Qaeda and Secretary of State Colin Powell specifically singled out Ansar when he laid out the U.S. case against Saddam Hussein to the United Nations on February 5, 2003.

According to Powell, Ansar was operating a "terrorist poison and explosive training centre" in the village of Sargat. The laboratory had been established by Abu Musaab al Zarqawi, the al-Qaeda chief in charge of the network's production of weapons of mass destruction, Powell alleged, and that was proof of a "sinister nexus" between Saddam and bin Laden's network.

Not only were American intelligence agencies keen to eliminate that threat but they also wanted to get into the camps to gather information that may be useful against al-Qaeda. Several months before the war, U.S. special operators had slipped into the area to link up with Peshmerga fighters and conduct an initial reconnaissance on the main Ansar bases.

The buildup of U.S. SOF in the area was no secret. In the weeks before the invasion started, journalists saw increasing numbers of American SOF and their CIA counterparts. There were also reports that American operatives had traveled to the town of Sulaimaniyah to interrogate Ansar prisoners being held by the Peshmerga as part of their planning for an assault on the group's bases. The covert operators didn't take kindly to any publicity. When journalists photographed and videotaped one of the teams, their material was confiscated.

Sporadic fighting between Ansar and Peshmerga forces started on Friday, March 21. But it wasn't until shortly after midnight on March 22 when full-scale American operations began with the firing of 40 Tomahawk cruise missiles into the Ansar camps. This was followed by bombing from B-52s flying from bases on the island of Diego Garcia in the Indian Ocean as well as carrier-based F-18 fighter jets. Much of the fighting was concentrated around the town of Khurmal, which had been seized by the 600-strong Ansar force. As thousands of the town's residents fled, Peshmerga and Ansar fighters fired back and forth at each other with artillery and small arms.

On Saturday, March 22, four transport aircraft carrying U.S. special operations forces landed at an airstrip just outside Sulaimaniyah while three busloads of American commandos were seen arriving in the town of Halabja the next day.

The arrival of an estimated 100 American SOF bolstered the 8,000-strong Peshmerga force and set the stage for a joint assault on the Ansar terrorist camps. For almost a week, the Ansar positions were hit by bombing raids and pinpoint attacks from Spectre gunships.

The main push on the camps came around 10 p.m. on March 28 when Green Beret A-Teams called in an aerial bombardment to cover their movement up the mountain slopes near the Ansar villages. Around 72 Green Berets as well as Peshmerga forces were involved in the attack, hitting the terrorists with 60 mm mortar fire and grenades from Mk-19 automatic launchers. Ansar forces responded by using anti-aircraft guns to lay down ground fire on the approaching Americans as well as launching Katyusha rockets.

A lone Ansar sniper was able to keep many of the Peshmerga pinned down, putting a stop to the advance for almost a day. U.S. special operations forces wanted to lob mortar rounds at the sniper but he was too close to Kurdish forces for that to succeed without risking Peshmerga casualties. The Peshmerga were eventually able to kill the Ansar gunman when they outflanked his position.

The U.S. SOF deployed their own sniper teams armed with .50 calibre McMillan and Barrett rifles. They patiently waited, hiding behind boulders on hillsides, for Ansar to appear from caves or along ridges high up in the mountains. Some of the American snipers dubbed their quarry "deer" as they fired over distances of a kilometre or more at Ansar hurrying along switchbacks. Occasionally, the snipers would spot for other special operations officers as they lobbed 60 mm mortar rounds at groups of Ansar fighters moving among the rocks. During a lull in the fighting, some Ansar would stand up along the ridgeline, taunting the Americans and Peshmerga with yells of "Allah Akubar" ("God is Great").

Fighting intensified when Ansar troops retreated beyond the snow line and occupied a series of caves and key mountain passes. Ferreting them out meant either starving the terrorists with a long protracted siege or calling in air strikes to collapse cave entrances. The air strike option was the one most often used. But in a situation similar to the battle at Afghanistan's Tora Bora in December 2001, many of the Ansar leaders apparently managed to escape. They hiked through the mountain passes into Iran, while a rear guard willing to be martyred fought the Green Berets and Peshmerga troops.

The fighting ended on March 31, with special operations soldiers counting more than 130 enemy dead. American officials played down suggestions that many Ansar guerrillas had escaped. Special operations spokesman Major Tim Nye acknowledged that a few had made it over the mountains into Iran but said that most of the force was destroyed by Kurdish and American firepower.

U.S. troops clad in chemical warfare suits sifted through the remains of the Ansar bases and special operations officers claimed that they had found evidence at Sargat that showed al-Qaeda was involved in developing chemical or biological weapons there. Materials seized at the base were sent back to the U.S. for further testing but nothing concrete ever came from those discoveries.

The building that Powell had described to the UN as a poison factory had been totally destroyed by artillery and cruise missiles. Shortly after the secretary of state made his allegations against Ansar, the group had allowed foreign journalists to tour the site and, according to those individuals, the building appeared to be a video- and audio-recording center used to create propaganda tapes. When journalists once again toured the site, this time after it was overrun by U.S. and Kurdish forces, they could see smashed circuit boards and loudspeakers in the ruins of the building.

U.S. Defense Secretary Rumsfeld later hinted that chemical and biological weapons may have been taken out of the Ansar villages before the attack, noting that earlier surveillance of the area indicated there had been a large number of trucks moving people and equipment out of the region.

While the attacks on Ansar may not have provided the conclusive evidence the U.S. wanted about an al-Qaeda-Saddam link, the decision to team up special operations forces with the Peshmerga was paying off. The drive through the north was rapid and Iraqi forces were soon on the retreat.

The successes, however, were not to come without casualties. On April 7, a column of vehicles led by U.S. SOF and Waji Barzani, commander of the Kurdish special forces, was moving towards the town of Dibajan, 36 kilometres southeast of Mosul. Dibajan had just been captured by Kurdish troops but Iraqi forces were still scattered between the town and the advancing Kurds and Americans.

As the operators were moving up the road, they could see at least one Iraqi tank moving about a kilometre and a half away. The decision was quickly made to call in an air strike on the armored vehicle and almost immediately two F-14 Tomcats responded to the mission. The U.S. operators and Kurds stood by their vehicles, which had been marked with bright orange panels to distinguish them as friendly forces for coalition aircraft. Nevertheless, one of the F-14s came in low over the column and dropped a single bomb on the Americans and their Kurd allies.

Vehicles were flipped over or burst into flames from the massive explosion. Ammunition on board some of the trucks ignited, sending rounds flying into the air. A member of a BBC TV crew, following behind the SOF column, had been on a satellite phone to his mother in England describing how the F-14s were coming in low for an attack on the tank. He had just finished telling her the roar of the jet engines was the "sound of freedom" when the bomb hit. Eighteen Kurdish fighters were killed and another 45 wounded. Three Americans were also injured.

Other special operations groups had some close calls. At the end of March, a group of 40 Special Boat Service commandos was inserted by Chinook helicopter into the Mosul area with a mission to carry out reconnaissance and hit-and-run attacks near the town of Baaj. The British operators unloaded their Land Rovers, motorcycles and all-terrain vehicles from the chopper and then split up into smaller sections of 10 men to conduct patrols. One of those ran into an Iraqi ambush and, outgunned and outnumbered, the British commandos quickly called for a Chinook to extract the unit. Although they beat a hasty retreat and quickly boarded the chopper, the SBS men realized that two of their comrades had become separated from the main group during the ambush.

An al-Jazeera television broadcast on March 31 showed jubilant Iraqis driving around Mosul in one of the SBS Land Rovers. Iraqi officials also displayed captured equipment including 40 mm grenades, machineguns, radio equipment and the SBS all-terrain vehicle.

Iraq's information minister, Mohammed Saeed al-Sahhaf, described the incident as a significant defeat and said as many as 10 British SAS had been killed in the ambush. "Amazingly the Americans have pushed the British to do that," he told the Qatar-based network. "They pushed them ahead as an experiment. It is very tragic for the British."

The images and resulting media questions back in England forced the British Ministry of Defense to break its traditional official silence on special operations and deny that there were any such fatalities. It acknowledged that British soldiers had been "extracted" from northern Iraq and that equipment had been lost. It did not identify the unit and did not mention the two missing commandos.

Meanwhile, the two SBS men began their escape and evasion strategy, setting off for the Syrian border 100 kilometres away. Although Syria might not provide the most friendly welcome, it would at least be better than the two would receive if they fell into Iraqi hands. During the day, the SBS men hid among the rocks and under brush only emerging at night to continue their trek. They eventually crossed into Syria where they were promptly arrested and, later, after some diplomatic maneuvering, were turned over to British authorities.

In the south, while conventional troops continued their drive towards Baghdad, special forces operators focused on seizing targets that might be used in the defence of that city. Of particular concern were several dams north of the capital which General Franks worried might be blown up in an attempt to slow down or stop his push.

On April 2, a team from Task Force 20 struck at the Haditha Dam, 200 kilometres northwest of Baghdad, running into heavy resistance from Iraqi troops. It took three days for the team's Delta Force operators and Army Rangers to gain control of the dam, which held back the waters of the Euphrates River, but after a thorough search for explosives, the commandos found no evidence to indicate the Iraqis had intended to blow up the facility.

Ninety kilometres northeast of Baghdad, a combined force of several dozen SEALs and Polish GROM, flown in from Kuwait, seized the Mukarayin Dam and power station. Pilots with the U.S. Air Force's special operations command earned their pay that night as they carefully positioned their Pave Lows over the dam while avoiding dozens of high-voltage hydro towers and power lines. The teams fast-roped from the choppers and quickly began working their way to the dam's control station and other key valve systems which controlled the water levels. The insertion went off relatively smoothly, except for a GROM operator who broke his leg during the fast-rope.

The Iraqis who operated the power station and dam were surprised at the appearance of the commandos, but they did not offer any resistance. Five hours later, a thorough search by the SEALs and GROM determined that the Mukarayin had not been rigged by the Iraqis for destruction. Even so, the assault team was ordered to hold the dam and hydro plant in case the Iraqis tried a counter-attack. They would stay at the site for another five days.

Around the same time, special operations troops started to move into Baghdad in greater numbers. There had been earlier reports that CIA and Delta Force operators had slipped into the capital in the early days of the war to try to target the Iraqi leadership, but now U.S. troops had a foothold in the city's suburbs and had occupied its airport.

In early April, Australian SAS Major Nick Withycombe accompanied U.S. SOF as they landed in an MC-130 at Baghdad's airport. It wasn't exactly a smooth approach, as Iraqi troops located near the airport peppered the lumbering plane with gunfire. As the coalition commandos tried to secure the airfield, fierce firefights continued and the Iraqis were able to mount counter-attacks on several occasions. Although Withycombe had expected coalition SOF to encounter more resistance, he did credit the Iraqis with being well-organized. One thing he noticed that indicated the Iraqis were a disciplined force was that they took their dead and wounded from the battlefield as they withdrew from the fight.

As they had done since the beginning of the war, Delta Force, the British SAS and CIA paramilitary operatives continued their search for Iraqi "leadership targets." It was extremely dangerous work, with operations at times taking place in the heart of the capital city. But it was deemed to be well worth the effort. If a special forces operator could take out Saddam or his sons with a well-directed sniper shot or a smart bomb, there was a good chance the Iraqi war effort would collapse.

Operators raided one of Saddam's palaces at Thar, 80 kilometres northeast of Baghdad, but it was discovered empty and there were no useful leads gathered there about the Iraqi leader's whereabouts.

Helping SOF in their search was Grey Fox, the secretive signals intelligence collection group. Officially known as the Intelligence Support Activity, the unit had been set up in 1981 and given the job of covert missions in the field to intercept all forms of communications. Grey Fox had worked in Bosnia, helping track down war criminals and was instrumental in the Colombian government-Delta Force mission to hunt down drug lord Pablo Escobar. Now, it was trying to pin down any communications among members of the Iraqi leadership, in particular Saddam Hussein and his sons, Uday and Qusay.

To do that, Grey Fox operators were flying unmarked signals interception aircraft over the regions around Baghdad, scanning for any radio or satellite phone transmissions. Strangely, however, there was only silence.

Some of the special operations teams, however, had some success in tracking leadership targets, or so it was first thought. On April 5 in Basra, the British SAS were credited with helping kill the infamous "Chemical" Ali Hassan al-Majid, who had been responsible for using chemical weapons against the Kurds in the 1980s. The SAS had been staking out a three-storey building in the central section of the city after receiving a tip that al-Majid, the Iraqi commander in the south, was inside meeting with senior officers. Using laser designators, the SAS marked the target and called in an air strike by a Harrier jet. Though the bomb hit the building, it failed to explode because of a faulty fuse. A tense nine to 10 minutes passed before a second Harrier could reach the target and drop a bomb which collapsed the structure. Artillery strikes were also called in on the office by the SAS. (The credit for killing al-Majid was premature. He was captured very much alive in Iraq almost five months later.)

Despite the ongoing fighting in Basra, British special operations forces were able to covertly move about the city, at least most of the time. In early April, *Daily Telegraph* journalist Olga Craig bumped into two special forces operators on a bridge, at which point one promptly greeted her with, "Who the fuck are you?" Dressed in civilian clothes and wearing T-shirts, the men would only tell the journalist that they were SOF being used in support of the British armored division in the city and that their job was to gather intelligence from a network of Iraqis inside Basra. Although Craig assumed the men were SAS, they could also have been SBS.

By April 10, most of the Iraqi leadership had vanished and an intense hunt began throughout the countryside. Operators set up checkpoints on the main roads between Baghdad and Tikrit on the possibility that Saddam might try to escape to his traditional stronghold.

Interdiction teams also lay in wait along escape routes to Syria near Qaim and Mosul. On several occasions, they stopped busloads of non-Iraqi fighters who were trying to get into the country from Syria to lend their support to Saddam, seizing weapons and taking the men into custody.

It wasn't long before the interdiction teams got results. On Friday, April 11, an Australian SAS patrol captured 59 Iraqis in a convoy as they were trying to leave the country. Most appeared to be Fedayen guerrillas and Baath Party members.

As they searched the bus and two cars, the SAS discovered weapons, gas masks, radio equipment and $600,000 (U.S.) in cash. Documents were also discovered which offered a reward of $5,000 for anyone killing a U.S. soldier.

The SAS troops continually changed their locations so as not to set up a recognizable pattern, thereby retaining the element of surprise. "Clearly, if you set a pattern, this gives the opposition the opportunity to prepare themselves with some sort of suicide bomb capability," explained Australian special operations command Chief of Staff Colonel John Mansell.

On Friday, April 18, a six-man British SAS patrol captured Saddam's half-brother, Watban Ibrahim Hasan al-Tikriti, as he tried to cross the Syrian border near the town of Rabia. He had been travelling down the highway from the northern city of Mosul when he was grabbed by the SAS, who had been tipped off by the Kurdish Peshmerga that Watban would attempt an escape along that route. Although he was no longer a ranking official, having been removed from power after a falling out with Saddam's son, Uday, his capture showed that special operations forces were serious about tracking down anyone associated with the regime who could provide information about the whereabouts of its leadership.

The mystery of what happened to some of Saddam's air force was also solved on April 16 when a force of 200 Australian SAS and commandos from the 4th Battalion, Royal Australian Regiment, seized the al Asad airbase 175 kilometres northwest of Baghdad. Before the war, military analysts had voiced concern that Iraq still had a fleet of 100 to 300 Russian-made fighter aircraft that could play an effective role in striking back at any invasion force. Strangely, the air force never took to the skies.

As the SAS approached the al Asad airbase, they found it occupied and defended by about 100 armed looters, who weren't inclined to give up their booty too easily. Some of the thieves manned an anti-aircraft gun and fired several rounds at the Australians before a few well-placed shots by SAS snipers, combined with the sight of Royal Australian Air Force F/A-18 fighter jets flying cover overhead, prompted the looters' quick retreat.

As the troops fanned out, they discovered not only how large the airbase was on the surface, but that it had an extensive underground tunnel system where protective biological and chemical warfare equipment – suits and antidotes – was stored. In the base hospital, the SAS discovered fresh bloodied field dressings on the floor, indicating the medical staff had recently fled. Officers' uniforms still hung on racks in their barracks. A giant mural of Saddam and fighter jets rocketing into the sky dominated the outside of the headquarters building.

Not knowing if there were booby-traps or mines among the buildings, the troops spent a cautious 36 hours going room-by-room to clear the facility.

The biggest find was discovered when the SAS started searching the hangers and the area surrounding the airbase. Fifty-one fighter aircraft, mainly older MiG-21s but also three modern MiG-25s, were found. Some of the planes just sat abandoned in hangers, but many of the aircraft were covered in camouflage netting or hidden in waddies. Many were still operational. Troops also discovered a French-made Roland air defence missile system and almost 8 million kilograms of ordnance, including missiles, bombs and anti-aircraft shells.

(After the war, more MiG-25s and Su-25 aircraft would be found buried at al-Taqqadum air field west of Baghdad. One of the MiG-25s had been completely covered in sand except for its tail fins which stuck out of the ground.)

Coalition commanders explained why Iraq's air force was noticeably absent from the war – its generals had been given an offer they couldn't refuse: if their planes flew, they would be destroyed. "We established means of contacting some senior Iraqis and we also sent them some personal messages saying: You really don't want to do this," said Brigadier Maurie McNarn, commander of the Australian forces in the Middle East.

Although the Australians had found much of Saddam's air force, Task Force 20 would come up empty-handed on its main mission – to find the Iraqis' alleged stockpiles of weapons of mass destruction.

The task force had been equipped with all the gear it needed to track down WMDs. It had portable laboratories for the analysis of biological and chemical agents. Its strike force, outfitted with MH-47 Chinooks, MH-60 and Little Bird choppers, could be rolling on a target within 60 minutes of getting the order to move. The task force, made up primarily of Delta Force operators and SEALs with support from U.S. Army Rangers, had been able to secure many of the suspected WMD target sites throughout Iraq, shipping hundreds of samples back to laboratories in the U.S.

But many of its efforts played out in a similar fashion to its first operation near the beginning of the war when it hit a suspected Scud base near Qaim. Task Force 20 operators quickly killed the Iraqi guards at the installation but no Scuds were found at the site. A search of the desert base did find several boxes of what appeared to be landmines designed to hold chemical or biological agents. The weapons were immediately sent back to the U.S. for testing but the results were less than conclusive. While the landmines may at one time have held a biological agent, the contents of the weapons were so old and deteriorated that it was impossible to give a definite answer as to whether these were WMDs.

The task force was also able to take into custody several high-profile Iraqi weapons scientists, including Rihab Rashid Taha, nicknamed "Dr. Germ," and Huda Salih Mahdi Ammash, the infamous microbiologist known as "Mrs. Anthrax." But under interrogation, the scientists revealed little useful intelligence on Saddam's WMD program.

Task Force 20's most public and controversial mission centered around its April 1 rescue of U.S. Army Private Jessica Lynch from a hospital in Nasiriyah. One of the task force's secondary missions was to rescue coalition POWs and its raid on the hospital would initially be greeted as a spectacular show of American resolve and SOF might. A little more than a month later, however, the raid was mired in allegations that it was simply a propaganda mission designed to boost morale on the home-front at a time when the U.S. war effort appeared stalled.

Lynch had been captured on March 23 when the maintenance unit she belonged to lost its way and was then ambushed by Iraqi forces. Nine of her comrades were killed and another five captured. The 19 year old was separated from the other captives and taken to Saddam Hospital in Nasiriyah for treatment of her injuries.

Shortly after midnight on April 1, the raid was launched with an assault force of Army Rangers and SEALs landing at the hospital in an unlit Black Hawk helicopter. Fifteen minutes before the commandos landed, troops knocked out Nasiriyah's main power station, plunging the city into darkness. A second diversionary attack hit a pocket of Iraqi militia while Harrier jets bombed the local Baath party headquarters.

Outside the hospital, the Rangers set up a perimeter while a SEAL team moved inside, going room to room in their search for Lynch. Overhead, a Spectre gunship orbited in case it was needed for fire support.

"Go, go, go!" the SEALs yelled as they stormed into one of the operating rooms. Doctors and medical staff were handcuffed with plastic restraints.

Lynch was finally found upstairs in bed with a nurse sitting by her side. "Private Lynch, we're U.S. soldiers here to rescue you," one of the men said to her.

The SEALs put the soldier on a stretcher and loaded her on an awaiting chopper. The total time spent in the hospital during the raid was five to 10 minutes and the entire mission was videotaped through night-vision equipment and transmitted live so that General Tommy Franks could watch the assault unfold.

The U.S. military wasted no time in providing journalists with details of the raid, which they described as the first successful rescue mission of a U.S. POW since the Second World War. "It was a classic joint operation done by some of our nation's finest warriors ... loyal to the creed they know that they'll never leave a fallen comrade behind and never embarrass their country," Brigadier General Vincent Brooks proudly told reporters.

Through leaks to news media outlets, U.S. government officials painted a picture of Lynch as a hero. Journalists were told how the petite soldier had fiercely fought off Iraqi troops before being injured in battle. Suffering from multiple gunshot wounds, she continued firing her M16 until it ran out of ammunition. At the hospital, she had been mistreated, slapped by a "man in black" – one of Saddam's intelligence officials – as he tried to question her. Or so the story went.

In fact, her "multiple gunshot wounds" turned out to be fractures suffered when her vehicle crashed into an American Hummer during the ambush. She couldn't remember anything about the attack. Journalists from the *Toronto Star* and the BBC interviewed medical staff who cared for Lynch and were in the hospital at the time of the raid. They told a very different story than the Pentagon's version.

Lynch wasn't abused by the Iraqis and hospital staff went out of their way to care for her, using up scarce blood and medical supplies to ensure the American received the best care possible. In fact, the hospital staff had tried on at least one occasion to bring Lynch to the American lines. Risking their own lives to drive through Iraqi roadblocks, they transported the young private to a U.S. Army checkpoint. But the ambulance driver panicked when he heard gunfire, and although he wasn't certain it was coming from the checkpoint, he decided to quickly turn the vehicle around and head back to the hospital.

The day before the raid, Iraqi soldiers who had been at the hospital cleared out, leaving most of their weapons and explosives behind. One of the doctors at the hospital, Hazem Rikabi, later said that if U.S. military forces had simply asked for Lynch, the hospital staff would have happily handed her over. The dramatic mission, he concluded, was to make the commandos into heroes. Another man, who lived beside the hospital, said he told an interpreter who was with the commandos that there were no soldiers inside, but they stormed in anyway. A BBC TV report called the raid "one of the most stunning pieces of news management yet conceived."

Predictably, members of the special operations community reacted angrily to the criticism and suggestions that the raid was staged. They rightly pointed out that they didn't know what they would encounter in the hospital and had to go into the building on the assumption they would be met by Iraqi troops who had no intention of surrendering Lynch.

It was obvious the special operations troops did their job. What remains uncertain, however, is whether the U.S. government attempted to manipulate the Lynch story into more than it was.

Two weeks after the rescue, major fighting in Iraq would be declared over. In the months ahead, U.S. troops would face a different type of war, more of a guerrilla conflict as Saddam's loyalists continued to fight in hit-and-run attacks.

Although details about special operations casualties during the Iraq war remain clouded in secrecy, it is believed they were low. Among the foreign coalition operators, both GROM and the Australian SAS confirmed that they had no casualties.

Not only did GROM operators escape unscathed, but their Iraq mission may have salvaged their 300-member unit from being dismantled or weakened by Polish military officials who weren't keen on SOF. "This war saved GROM," the unit's founder, retired General Stanaslov Petelicki told *The Weekly Standard*. "Without it, it would have been broken up between the army and navy. But now everyone knows about GROM in Poland and they are proud of them."

For its part, the Pentagon acknowledged several injuries, such as those of two operators seriously wounded on the outskirts of Baghdad on April 7. Two combat search-and-rescue helicopters flew through a sandstorm to pick up the injured men just south of the Iraqi capital. Once again, special operations helicopter crews proved their mettle, piloting their HH-60G Pave Hawks through the sand and dust which cut down visibility to less than a kilometre.

The two choppers were backed up by four A-10s, a refueling aircraft and a Special Operations MC-130E Combat Talon which had a flight surgeon and medical staff on board. The crew successfully flew into a hot landing zone, where the men were taken on board and then flown to an airstrip at Najaf, 90 kilometres south of Baghdad. There, they were transferred to the Combat Talon and flown to a hospital in Kuwait.

The Pentagon released the name of one special operations airman, Staff Sergeant Scott Sather, who was killed in combat in southern Baghdad on April 8. It did not give any details except to say that the 29 year old was assigned to the 24th Special Tactics Squadron out of Pope Air Force Base, North Carolina.

CENTCOM commander General Tommy Franks, once part of the ranks of those in the regular army who were suspicious of special operations forces, had now seen what such units could do during two wars that he had planned and executed. He had flooded western Iraq with commandos to hit the country's missile bases. He had created a second front in the north using mainly Green Berets and Air Force combat controllers. In fact, a U.S. Air Force study released a month after the war showed how critical special operations forces had become. The small number of operators on the ground – making up about one-tenth of the invasion force – were responsible for providing a quarter of all targets for aircraft to attack.

The general had only the highest praise for the special operations units. "They have accomplished some wonderful things out there," he said.

TOP: *Map of Iraq.*

LEFT: *U.S. Air Force pararescuemen and officers check their equipment before deploying to an undisclosed location during the war in Iraq. (COURTESY USAF)*

ABOVE: *SEALs on patrol during Operation Iraqi Freedom.* (COURTESY USN)

RIGHT: *SEAL team members practice seizing ships and oil rigs in preparation for the war with Iraq.* (COURTESY USN)

OPPOSITE PAGE, TOP LEFT: *SEALs practice fast-roping in preparation for their missions during Operation Iraqi Freedom to seize oil pumping stations and rigs.* (COURTESY USN)

BELOW: A Royal Marine commando in action on the Al-Faw peninsula. The Marines worked with Polish GROM and U.S. SEALs during that operation. (COURTESY MOD)

BOTTOM: A special operations helicopter crew member gets ready for a mission into Iraq. (COURTESY USAF)

RIGHT: An Australian SAS patrol stops to talk to Iraqi civilians in an effort to gather intelligence and win hearts and minds.

BELOW: Australian SAS long-range patrol vehicles temporarily stopped in Iraq's western desert.
(BOTH COURTESY ADF)

BOTTOM: SEALs seize a boat at the port of Umm Qasr. (COURTESY USN)

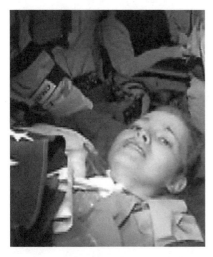

An SAS patrol outside Al Asad airbase in northwest Iraq. The Euphrates River is in the background. (COURTESY ADF)

TOP: *Some of the remains of Iraq's air force after being destroyed by SOF raids. (COURTESY ADF)*

MIDDLE LEFT: *U.S. Army Private Jessica Lynch being carried to a helicopter by SOF members of Task Force 20. (COURTESY USN)*

LEFT: *Coalition troops examine Iraqi sea mines discovered by SOF and naval forces. (COURTESY USN)*

TOP: *Australian troops from 4RAR (Commando) and the SAS pass by camouflaged Iraqi aircraft at the Al Asad airbase.* (COURTESY ADF)

ABOVE: *A U.S. SOF, presumably a SEAL, takes cover while at the Iraqi port of Umm Qasr.* (COURTESY USN)

OPPOSITE PAGE: *A Pave Hawk helicopter hovers over a U.S. special operations soldier during an infiltration training exercise. The Bush administration has announced a significant increase in the number of SOF soldiers over the next several years.* (COURTESY USAF)

Expansion and the Four Truths

Over the years in the SOF community, and especially in the U.S., a set of rules has been developed and refined as a result of lessons learned in previous wars. These rules are generally referred to as the "Four Truths":

- Humans are more important than hardware.
- Quality is better than quantity.
- Special operations forces cannot be mass-produced.
- Competent special operations forces cannot be created after emergencies occur.

In the two years following the September 11 attacks, some in the military community have been concerned that the Pentagon and Bush administration are making changes that will lead to the breaking of the last three rules. Such critics warn that the end result of disregarding these truths could be a less competent and less effective special operations soldier.

U.S. Defense Secretary Donald Rumsfeld has called for a relatively modest expansion of SOF over the next several years. The increase in numbers being discussed has ranged from 4,000 to 9,000 new operators. The concern with such a boost in numbers, however, is that in the rush to produce more special operations troops, training standards will drop substantially.

At the heart of the problem is the long-standing shortage of qualified recruits coupled with the considerable increase in the operational tempo for SOF because of the war on terrorism. In total, U.S. Special Operations Command has about 47,000 troops, although that number also includes support personnel, psychological operations and civil affairs staff, the pilots and crews who operate SOF aircraft, and units such as the U.S. Army Rangers, which are made up of younger and less experienced soldiers.

In reality, the number of SOF "trigger-pullers" – those doing the actual fighting – is less than 10,000.

Even before the al-Qaeda attacks on New York and Washington, special operations units were starting to feel stretched. The war in Kosovo and missions in Bosnia put Delta Force, the SEALs and Air Force combat controllers on the ground to help in special reconnaissance and target selection. Some of the units, such as Delta and the SEALs, were also engaged in the hunt for war criminals. At the same time, there were ongoing lower-profile SOF missions in more than a dozen countries.

With the wars in Iraq and Afghanistan, almost half of SOCOM's personnel found themselves overseas. And, unlike previous missions, such as the invasions of Panama and Grenada, these were not short-term deployments.

The number of personnel currently serving in the U.S. military – the recruiting pool from which special operations draws – underlines the extent of SOCOM's problem. During the Cold War, American forces numbered almost 2.5 million soldiers, sailors and aviators, enough to maintain SOF recruiting for that period. By 2003, the recruiting pool had shrunk to 1.4 million personnel while the need for SOF had skyrocketed.

The result has been a shortage of skilled operators. The Navy estimates it needs to recruit 250 SEALs a year to meet its needs and deal with attrition, but it has only been able to select 200 annually. The Green Berets are supposed to be able to field 270 A-Teams with 12 men each; instead, in 2003, they had operators for only 225 teams.

While the demand for SOF to be on the front-lines of the war on terror continues to grow, it still takes a great deal of time and money to recruit and train such operators. On average, a Green Beret needs two years of training before he is fully qualified. It takes about three years to produce a combat-ready SEAL, and just to train such an operator in his first year costs an estimated $800,000 (U.S.).

Despite such problems, the further expansion of SOF units is already underway.

For instance, in June 2003, Admiral Vern Clark, the chief of naval operations, announced that he intended "to grow" the SEALs over the next several years. No figures were released officially, but some Navy officers suggested that within five years another 272 SEALs would be added, enough for two new SEAL teams.

Earlier, in 2002, the U.S. Army devised a plan for increasing the size of the 160th Special Operations Aviation Regiment from 1,800 to 2,700 people over a five-year period.

As well, the U.S. Marine Corps announced that it will create an 86-man SOF team for use by SOCOM. The unit would deploy with SEALs and be used to augment the naval warfare teams in reconnaissance missions and direct action operations. This move is an about-face for the Marines, which was the only service that didn't put units under the SOCOM umbrella when that command was created in 1987. At the time, there was a reluctance to turn over control of Marines to another organization. So instead, the Corps determined that special operations skills would be among the capabilities of its force reconnaissance teams as well as Marine units which were given a "Special Operations Capable" label. The war in Afghanistan, and the need to provide SOCOM with more support, changed that way of thinking.

The CIA has also started to expand its paramilitary capabilities by creating new teams to hunt terrorists overseas. Recruitment for those units could put even more pressure on SOCOM, as serving and former SOF military personnel would be the most obvious candidates for such positions. In fact, the first CIA team into Afghanistan on September 26, 2001, included a former SEAL.

Faced with a smaller recruiting base, how will such units be expanded or even maintained? There are several ways.

One is the temporary introduction of "stop-loss" programs, which prevent SOF personnel from leaving their units or military service as long as such orders are in place. In October 2001, the Air Force and Navy announced their stop-loss programs. On December 4 of the same year, the Army brought in its program, freezing in place almost 1,000 SOF personnel. Included in those numbers were almost all categories of Green Berets, as well as pilots for MH-47, UH-60, MH-60, CH-47D and OH-6 helicopters. The last time the Army used the stop-loss program was during Desert Storm in 1991.

The other more long-term solution, however, is to significantly increase the number of operators being recruited and trained. But along with the push to produce more special operations troops comes the concern that training standards will suffer.

Maintaining high SOF standards has never been easy, even before the decision to significantly increase their numbers. These fighters, after all, are expected to be among the most lethal, most intelligent and most effective in the world. Besides being physically fit, most operators are veteran soldiers who bring a certain maturity and confidence to the battlefield. For example, a typical Green Beret is a sergeant with 10 years' experience in the Army. Most are in their 30s and are married with children.

To increase their pool of recruits, the Green Berets are now taking civilians straight off the street. The Army has received a strong response to the SOF recruitment program from civilians. Candidates have been drawn to the Green Berets both because of the supposed glamour of the job and because of positive media coverage of the unit during the wars in Afghanistan and Iraq.

Civilian recruits undergo nine weeks of basic training followed by two years of specialized combat skills. Although they initially enter the Army as privates, they earn the rank of sergeant upon completion of their two-year Green Beret program.

Army officials claim that off-the-street recruitment, dubbed the Special Forces Recruiting Initiative, is not linked to the events of September 11. In fact, they argue that it is a return to the roots of the special operations recruitment process of the 1950s which allowed civilians, in addition to Army personnel, to be selected. According to the Army, the recruiting initiative will not mean lower standards, but will actually produce a better soldier. Under the new process, the Green Berets carefully select the candidate off the street and then mold him as a special operations soldier right from the beginning of his career.

Army special forces aren't the only ones changing how they recruit and train personnel. Candidates for Air Force combat controller positions used to be put through an intensely physically demanding four-month selection process which had a high failure rate. As a result, the Air Force was continually short of such operators. There were also concerns that potentially good candidates were being weeded out of the selection process early on since they couldn't meet the initial physical fitness requirements or had been injured during the selection process.

A new training regime, instituted in 2001, saw recruits mentored by active duty or retired combat controllers. In addition, the training regime was spread out over time. Instead of 12 weeks of gruelling physical activity, there is a two-week indoctrination course followed by 15 weeks of instruction in air traffic control. This is followed by another three weeks of survival school and four months of combat control school. After that comes a 12-month Advanced Skills Training Course in which the recruits are mentored by experienced Air Force SOF to bring them up to combat-ready status.

"We no longer expect these young men to go from 'zero to hero' in 12 short weeks," Colonel Craig Rith, commander of the 720th Special Tactics Group, explained to *Armed Forces Journal International* in September 2002. "Now we work with them and move them to a higher stage gradually, spreading out the expected level of proficiency over a longer period of time."

The failure rate for candidates, which was previously 85 per cent, has been cut to half that.

In addition to calls for maintaining high recruiting and selection standards, supporters of the special operations community have also raised warnings about the overuse of SOF in the future. David Litt, a former ambassador and a SOCOM advisor, has recommended that the U.S. government only give the most essential missions to SOF. He notes that there is a need for greater use of intelligence gathering, diplomacy and civilian law enforcement in the war on terrorism.

Others have also recommended that regular force units take over some of the roles now assigned to SOF, such as combat search and rescue and NEO (non-combatant evacuation operations), which include rescuing civilians trapped in war-torn countries. In addition, they argue, special operations forces do not necessarily need to train foreign armies or be involved in de-mining efforts; those missions could be handled by regular U.S. Army or Marine units. As well, SEALs shouldn't have to continue doing almost all non-hostile ship boardings as that job could be conducted by regular sailors.

In January 2003, Defense Secretary Rumsfeld brought in a series of changes that addressed some of these recommendations and appears, at least initially, to be a sound step in supporting special operations units.

His proposal includes transferring some current SOF duties, such as training foreign militaries, to conventional units. Those conventional forces would also be required to pick up the slack in areas such as airlift for SOF as well as taking over some counter-drug operations.

Rumsfeld's proposed changes are welcome news to the special operations community. For example, his directives would put an end to the use of SOF aircraft for flying non-essential missions. This was a problem that had arisen over the years because regular force commanders, who had control over SOF aircraft in the field, tended to use these highly skilled crews for difficult non-combat operations, such as flying in extremely poor weather. Although useful for crew training, such missions tied up SOF personnel and equipment which could have been used for other more important operations.

In a few cases, the changes that Rumsfeld announced in January were already being implemented. For instance, in late 2002, U.S. Marines began replacing Green Berets in a program to train soldiers in the former Soviet republic of Georgia. The program is an important one in the war on terror since the training is intended to enable Georgian forces to hunt down Islamic guerrillas who are infiltrating into the country from Chechnya.

The other major change that Rumsfeld ordered was to give SOCOM more flexibility in determining its missions, as well as the type of equipment and personnel it could call upon from conventional units. So, in the future, regular force commanders will sometimes find themselves being used to support SOCOM missions.

"What it means in practical terms is the theater special operations command would have access to... (regular force) units that would act in response to its direction and control," explained a senior SOF officer.

The ultimate effect of such changes remains to be seen. Will they lead to a more effective military force in the war against terrorism? Will Rumsfeld's decision to expand SOF have a detrimental effect on the quality of commandos being produced? Only time will tell.

Still, while U.S. defence officials clearly have their work cut out for them in expanding American SOF units, their counterparts in smaller allied armies, who don't have access to the same relatively large pools of funding and personnel, are faced with even greater challenges.

ABOVE: SEALs practice fast-roping onto one of the unit's patrol boats. SOF operators are in high demand to fight the war on terror. (COURTESY USN)

TOP: A U.S. Air Force combat controller jumps off the back of an MC-130E Combat Talon aircraft. The combat controllers have modified their training regime in an effort to gain more recruits.
ABOVE: Green Berets in Afghanistan train new recruits of the Afghan army. Such tasks in the future may be taken over by conventional forces. *(COURTESY U.S. ARMY)*

ABOVE: *A SEAL undergoes training in the Nevada desert. The Navy wants to create two new SEAL teams over the next several years.* (COURTESY USN)

OPPOSITE PAGE: *Maritime operations have become a key mission for JTF2. Here, JTF2 operators use Zodiac boats.* (COURTESY JTF2)

JTF2: From Counter-Terrorism to SOF

9 Several thousand metres over the Pacific Ocean, members of the JTF2 sniper team stood on the open ramp of a Hercules transport aircraft, staring at the shimmering expanse of water below them.

On board the plane, a Canadian Forces jumpmaster gave the men the ready signal and they shuffled closer to the edge of the ramp. Moments later, the commandos propelled themselves out of the Hercules in a free-fall formation.

As their parachutes opened and they floated towards the Pacific, U.S. Navy sailors stood by in a boat to pluck the sniper team out of the water. The JTF2 commandos were soon speeding towards the shore, where they would slip into the dense forest of Canada's West Coast and head to a pre-designated target for their training exercise.

Elsewhere, a JTF2 dive team had gone ashore for its mission. Its members were to make contact with another soldier playing the role of "partisan" and then safely escort him to a submarine that was waiting several kilometres offshore.

The training mission, which took place on Vancouver Island, British Columbia, in October 1997, illustrates the grey zone in which Canada's Joint Task Force Two has operated during much of its existence.

Officially, the unit's role has been to deal with domestic terrorism, in particular with incidents involving the rescue of hostages or the seizure of buildings, aircraft or ships by terrorists.

Yet many of its actual operations, as well as training scenarios, have included a healthy dose of SOF activities, particularly those designed for overseas deployments. Some in the unit have wanted to expand JTF2 into a full-fledged special operations force, while others have argued for keeping it small and concentrating on its original mandate of domestic counter-terrorism and hostage-rescue.

In August 2000 a JTF2 training officer outlined for Canadian military leaders the unit's philosophy: "Stay lean, stay mean, but most of all stay focused."

But the September 11 attacks on the U.S. – so close to Canada's border and so specifically aimed at its biggest and closest ally – would make that advice difficult to follow. JTF2 might continue to be mean and perhaps focused in the future, but it wouldn't stay lean for long. Furthermore, JTF2's long-held practice of secrecy, even in relation to the country's highest-ranking officials, would be held up to scrutiny as some questioned the unit's lack of political accountability.

Unlike the American SOF example, in which numbers were to be boosted by roughly 10 per cent following September 11, the Canadian government decided on a much more significant increase in its special operations and counter-terrorism capabilities. A little more than two months after the attacks on New York and the Pentagon, the Canadian government announced it would *double* the capacity of JTF2 by 2006, spending almost $119 million (Canadian) in the process. At the time of the announcement, the unit had 297 members, so it has generally been accepted that JTF2 will increase to about 600-strong by the target date (although that figure includes support personnel).

The order was a double-edged sword. On one hand, it provided a large influx of cash and the chance to purchase much needed equipment. It also made JTF2 a major player in the pecking order of the Canadian Forces. On the downside, some argued, it would lead to a reduction in the quality of the unit's operators and a watering down of training and selection standards. In short, the Canadian government's decision to double JTF2 broke almost all of the Four Truths.

Officially, Canadian Forces officers continue to insist that JTF2's expansion is proceeding at an excellent pace and its tough training and selection standards are being maintained. Appearing in a 2003 JTF2 information video, Deputy Chief of the Defence Staff Vice Admiral Greg Maddison states that there is no leeway for marginal candidates in this "battle-seasoned" and world-class unit. "Very few (candidates) are successful in meeting all of its very high standards in physical and mental fitness, intelligence and role-related battle tasks," Maddison declares.

However, JTF2's own figures and records suggest a different story.

Well before September 11 and the government's decision to double the size of the unit, JTF2 was having difficulty recruiting enough troops to sustain itself.

In the mid to late 1990s, the unit had been trying to expand but was running into trouble because of the dwindling Canadian Forces recruiting pool. In 1993, when JTF2 was created, there were 80,000 personnel in the Canadian military. By the end of the decade that had shrunk to 60,000.

In 1995, 90 candidates were invited to the unit's Dwyer Hill base for its gruelling selection process and, of those, only nine would graduate to become JTF2 "assaulters," the term used to designate its front-line operators. This 10- per-cent success rate was in line with the selection standards of some of the top counterterrorism units in the world. The JTF2 selection process, however, was not capable of producing the numbers needed for the future. "It had become obvious that the status quo could not produce the required amount of assaulters necessary to sustain growth for the unit," a senior JTF2 officer wrote at the time.

As a result, the six-day selection process was "revised" and recruits were given three-week preparatory training sessions to ready them in advance for the actual selection regime.

After that, the number of special operations assaulter graduates steadily increased. In 1998, there were 17, and in 1999, there were 19. A new support role within the unit – mobility operators – was also created. These operators were given the job of handling boats and vehicles to get JTF2 assault troops to their targets. Of the 12 selected for that new training in 1999, 11 graduated.

By September 2000, relatively high success rates for recruits were becoming common. Of the 78 who were invited to JTF2's Dwyer Hill base for the selection process that year, 23 passed the special operations assaulter course and eight graduated from the mobility operators course. By then, JTF2 had also started to bring women into the unit, although they were in support roles. One female corporal was in the unit's communications branch, while another, a master corporal, was in its intelligence section.

In order to get the numbers it needed, JTF2 officers had little choice but to change the unit's selection standards. By 2000, although officially on paper the Canadian Forces was still around 60,000-strong, it had only 52,000 active members. JTF2 drew its ranks mainly from the Army which by that time had dwindled to 18,000 and, of those, less than half were considered front-line combat troops.

At the same time, retaining trained JTF2 operators was becoming a challenge. With countless hours spent away from home on exercises, burnout was a potential problem. And although many loved their jobs, some of the commandos were being recruited by civilian police agencies where their tactical and security skills were in high demand.

In a presentation to senior officers in August 2000, a JTF2 training officer noted that over the next three to five years recruiting was expected to become more diffi-cult. The massive downsizing of the Canadian military in the 1990s had taken its toll. There was also a shortage of new recruits for the regular military units from which JTF2 drew on. The counter-terrorism unit would have to put major resources into recruiting and selection just to maintain its current levels. "In the future we will have to spend more to get less," the major noted.

A month after September 11, 2001, JTF2 officers met to discuss how to handle expanding the unit and the effect such expansion would have on personnel selec-tion. That year, the unit had posted one of its best rates ever for successful candi-dates. In 2001, 65 recruits were invited to Dwyer Hill to try out for the unit; 32 graduated to become special operations assaulters, a success rate of 50 per cent.

Even though regular force Canadian soldiers are generally recognized as being a cut above many of their foreign counterparts and would make good SOF candi-dates, in the world of special operations, a high success rate in selection isn't neces-sarily a positive sign. Some units, such as the Polish GROM, say they only take around three per cent of recruits who try out for the unit. Others, such as the British and Australian SAS, are estimated to accept about 10 to 20 per cent. The success rate of recruits in the German Kommando SpezialKraefte is between 25 and 30 percent. It is argued that such tough and exacting criteria that lead to these lower rates help ensure a highly competent operator.

Even so, JTF2 officers decided that further revisions were needed in how opera-tors were selected if the unit was to meet the government's expansion objectives. One change discussed was to accept recruits who had done well in army-oriented field skills, or "green" training, into JTF2 squadrons even if their counter-terrorism "black" skills were not up to speed. "If black skills are not satisfactory but green skills are achieved, the member will be employed in SQN(Squadron) to develop black skills," JTF2 officers wrote in a November 2001 report.

Around this time, it was also decided to try to convert a maximum number of mobility operators – who handled JTF2 boats and vehicles – into assaulters. Ac-cording to the officers, this could be achieved by lowering the selection quality of assaulters or by maintaining present standards and offering selection to low-scor-ing candidates. The conversion would produce 14 or 15 more candidates for the special operations assaulter course in 2002.

As well, the basic requirements for recruits was downgraded. In 1994, anyone who even wanted to be considered for JTF2 had to have a minimum of four years of service. Only those who were corporals or above were allowed to apply. The idea behind that criteria was that such soldiers would be more skilled and profession-ally mature.

By 2002, those requirements had changed. Candidates for JTF2 could include privates, and any military personnel who wished to apply needed only two years of service.

By early 2002, senior military officials were finally acknowledging, at least privately, the problems associated with JTF2's expansion. At a meeting in February, a group of generals, including army commander Lieutenant General Mike Jeffery, voiced concerns about whether the Army, Navy and Air Force could provide the needed recruits for JTF2. The problem, they said, could be particularly acute for the increasingly small Army since JTF2 was now drawing many of its personnel from those ranks.

In March, JTF2 was also acknowledging the obvious, albeit in classified reports. "The primary risk associated with the project is that recruiting and training may not be able to sustain the rate of expansion," a JTF2 officer admitted.

Furthermore, JTF2 was facing competition from other specialized units in the Canadian Forces, which were also finding themselves short of personnel. In 2003, the Canadian military's Search and Rescue Technicians (SAR Techs) went on a major recruiting drive to boost its falling numbers. Like JTF2, this unit was feeling the full effect of the severe personnel reductions that the Canadian Forces had endured throughout the 1990s. In 2002, only 26 candidates had applied to become search-and-rescue specialists, about 10 per cent of the applicants the unit had received a decade ago.

Like JTF2, the SAR Techs promoted its elite training regime. Recruitment material showed how SAR Techs put their lives on the line daily, saving sailors from sinking ships, pilots from downed aircraft and hikers missing in Canada's rugged wilderness. Unit personnel were trained in parachuting, scuba diving and mountain climbing skills. The added bonus was that search-and-rescue technicians didn't deploy overseas, providing a more stable family life than that offered by JTF2. Highlighting this was a smart recruiting ploy on the part of the SAR Techs since there was a growing concern among military personnel that joining JTF2 put a strain on families – so much that the counter-terrorism unit's Dwyer Hill base had been dubbed "Divorce U."

Still, what might stop some applicants from joining the SAR Tech ranks was that the unit's initial recruiting standards were higher than those of JTF2: the physical fitness testing was more demanding, the unit required four years of military service, and the minimum rank allowed to apply was corporal.

To deal with the ongoing problem of finding enough candidates for JTF2, Canada's military leaders have also considered using the Army's three parachute-qualified Light Infantry Battalions as "feeder" recruiting pools for the counter-terrorism unit. But that still doesn't deal with the main problem: the Canadian Forces, particularly the Army, simply does not have enough personnel to properly fuel the desired expansion of JTF2. The most obvious solution, according to some defence analysts, is to increase the ranks of the Army.

Others warn that the government's order to boost the size of JTF2 will simply shuffle some of the best troops in the Canadian Forces into one unit. Robert Farrelly, executive director of the Royal Canadian Military Institute, argues that it makes more sense that JTF2 be used only for domestic counter-terrorism operations and selected overseas missions such as reconnaissance and battlefield intelligence-gathering. For those roles, a 300-member force is enough, he says.

Despite ongoing recruitment and expansion problems, one thing JTF2 hasn't had to worry about during most of its existence is excessive political oversight of its activities. In fact, unlike the situation facing JTF2's counterparts in other countries, senior Canadian politicians and diplomats have seemed almost blissfully unaware of the unit's activities.

During the 1999 war in Kosovo, there was a brief flurry in the House of Commons about the kind of controls governing the use of JTF2 when an MP suggested, erroneously as it would turn out, that its commandos were conducting reconnaissance missions behind Serbian lines. In response, JTF2 advisor Major David Last wrote in a briefing note prepared for Defence Minister Art Eggleton that although the commando organization was different from other units in its training and tactics, the "process for approval and oversight of domestic and international deployment is essentially the same as for any other element of the CF (Canadian Forces)."

By the time the Afghan war had started, this process appeared to have changed. Claiming a need for operational security, the military brought in a series of strict rules that severely limited what senior government officials knew about JTF2 missions.

Jim Wright, an assistant deputy minister at the Department of Foreign Affairs, would later testify in front of Parliament that under this system Canada's top military leaders decided the type of information the defence minister would be given about JTF2 missions. Then it was up to the minister to decide whether he wanted to inform the prime minister. All transfer of JTF2 information between senior military leaders and the government would be verbal and nothing would be written down.

The first casualty of this command and control regime was Defence Minister Eggleton, who became confused about information he was verbally given when JTF2 took prisoners in Afghanistan on January 20, 2002. His failure to promptly inform Prime Minister Jean Chrétien about the mission later sparked a House of Commons inquiry and allegations that the government had tried to cover up the fact that Canadian commandos had taken prisoners.

Military officers said it took Eggleton three verbal briefings before he realized the full nature of the JTF2 mission. In fact, the defence minister credited a newspaper photograph of JTF2 operators with their captives for partly clarifying the situation for him. "The old saying that a picture is worth a thousand words was never truer," he explained. "I was realizing, for the first time, the full extent of the involvement for Canadian troops in that operation."

That may have been so, but realizing the importance of a mission based on photographs of JTF2 operators inadvertently published in the news media hardly constitutes a proper political oversight for SOF missions.

Even Prime Minister Chrétien appeared to have been left out of the information loop on basic facts such as whether JTF2 had entered the Afghan war. When asked in December 2001 whether the commandos were in Afghanistan, Chrétien replied that he didn't know.

That lack of political oversight was only highlighted by the February 2002 hearings into allegations that Eggleton had misled Parliament. A group of senior officials from the Privy Council Office, the top bureaucratic agency for the Canadian government which also oversees defence and intelligence issues, stated it wasn't necessary for their organization to know specifics about JTF2's overseas missions. Senior diplomats who formulated Canada's foreign affairs policy were also unaware of JTF2's activities overseas. All of the senior bureaucrats claimed such secrecy was needed to protect operational security of the unit.

Richard Fadden, the Privy Council Office's deputy clerk, showed how little senior government officials knew about JTF2 when he testified at the hearing that the Afghanistan mission was the first time the unit had deployed outside Canada on a military operation.

In his appearance before the same Parliamentary committee, Chief of the Defence Staff General Raymond Henault claimed that in the past JTF2 carried out only out domestic missions. "Any travel abroad was only for training purposes, exercises or consultations with other similar forces," he told Parliamentarians. "This is the first time the unit or parts of the unit are used in operations abroad."

Not quite.

JTF2's longest-running mission ever was Operation Dubonnet in Bosnia. Initiated on July 14, 1996, the mission was to provide Canadian commanders in the former Yugoslavia with JTF2 "technical assistance" teams who were to be on hand in case Canadian peacekeepers were taken hostage by the warring factions. In fact, it was a December 1994 incident in which 55 Canadians were held by Serbs in Bosnia that laid the groundwork for such a mission.

For Operation Dubonnet, which ran until 1999, the JTF2 teams conducted counter-terrorism site assessments, figuring out how vulnerable Canadian soldiers and their bases were to hostage-taking attempts. As well, "other tasks" were added to the mission. By May 1997, JTF2 had five-man teams rotating through the former Yugoslavia. An example of the "other tasks" that the commandos performed was to provide back-up for soldiers from the 2nd Battalion, Royal Canadian Regiment, when they descended on a Bosnian army barracks looking for illegal weapons on May 25, 1999.

In the fall of 1998, JTF2 had also begun preparations for what had the potential to become its largest mission ever: the full-scale deployment of the unit to rescue Canadian troops and civilians trapped in the Central African Republic.

That year, a 45-member regular force unit of Canadian soldiers had been assigned to a UN peacekeeping mission sent to the former French colony. The situation in the Central African Republic was tenuous at best. Rebel forces occupied part of the country and the current government had experienced three army mutinies in the past two years. The Canadians, who were mostly headquarters and communications personnel, lacked the firepower to protect themselves if the situation deteriorated.

That's where JTF2 was to come in. Code-named COP Zombie, the mission called for the commandos to come to the aid of the soldiers and Canadian civilians threatened by any resurgence in fighting in the Central African Republic. Members from the 1st Battalion of the Royal Canadian Regiment, as well as a Canadian Forces field surgical team, were to be airlifted by C-130 Hercules transport planes into the Central African Republic to support Zombie.

Colonel Barry MacLeod, JTF2's commanding officer at the time, along with two other officers, traveled to the republic in November 1998 to conduct a reconnaissance for the potential mission.

To prepare for Zombie, JTF2 flew to Fort Benning, Georgia, in January 1999 to conduct a full-scale rehearsal of the rescue plan. After that, the unit was put on standby, ready to move out within seven days. As tensions eased in the Central African Republic, the level of readiness was downgraded to 21 days. While the larger JTF2 unit was never required for the mission, JTF2 operators were sent to the Central African Republic throughout 1999 to act as advisors for the Canadian regular force commander there.

It appeared from his testimony to Parliament in 2002 that General Henault, Canada's top military leader, didn't consider Dubonnet or Zombie as "operations abroad." Parliamentarians didn't know about the missions because they had been kept secret.

The situation in Canada, in which key political leaders, from the minister of foreign affairs to the prime minister, were ignorant of JTF2's operations overseas, stood in stark contrast to what was happening in other nations.

In the U.S., President George W. Bush was personally briefed by a special operations major general on the missions being planned in the wake of the September 11 attacks. In Germany, the GSG-9 counter-terrorism unit reported directly to the country's chancellor. The commanding officer of the British SAS had access to that country's prime minister. Britain's Foreign Secretary is also briefed on SAS missions overseas on the understanding that the unit's actions could affect the nation's foreign policy.

In Australia, Prime Minister John Howard has taken a personal interest in SOF, visiting SAS troops as they deployed and returned from Afghanistan and Iraq. In fact, like the Chrétien government, Howard has ordered a significant increase in the numbers of the country's special operations and counter-terrorism forces. Although this expansion may face some of the same problems encountered by JTF2, some observers believe the Australians are in a much better position to deal with such challenges partly because Australia already has an established 600-to 700-strong SAS, an air support structure that includes Chinook and Black Hawk helicopters, an established feeder unit in the guise of the 4th Battalion, Royal Australian Regiment, as well as a larger army.

Despite potential problems that might materialize in the future regarding political oversight and quality of recruits, for the moment JTF2 is riding high. As one JTF2 officer noted about special operations in Canada: "This is a growth industry."

Unlike other Canadian Forces units that are starved for cash, JTF2 has almost *carte blanche* on the purchase of new equipment. Its Dwyer Hill base is in the process of being significantly expanded.

For the Canadian government, JTF2 has become the almost-perfect solution to help deflect criticism that it hasn't been spending enough on defence. While neglecting the rest of the Canadian military, the government has concentrated on building up one small component that could be used on a regular basis on overseas operations. What's more, with the intense secrecy surrounding JTF2, the government can pick and choose which details it wants to release to the media and public. News that even a small team of JTF2 operators is deploying to a particular region is enough to gain positive media coverage of "elite" commandos heading off to war.

Speaking at a March 2002 conference on special operations at Canada's Royal Military College, Major B.J. Brister noted that among the advantages to the Canadian government of having a special forces unit are low cost, low risk and "deniability."

He argued that a nation with few military assets could make a significant contribution with a relatively small outlay in terms of personnel and cost. "An effective SF force can offer the advantages of flexibility, versatility, deniability, rapid deployment, low cost and low risk to a government and a people that are simultaneously averse to high defence budgets yet intent on maintaining their sovereignty and their place among nations," Brister noted.

"The casualty figures for a Special Forces organization may well be more politically acceptable (lower) than for a larger conventional military contribution," he added.

The angle of low cost and low numbers of personnel was also being pushed within government by JTF2 personnel. In its October 2001 expansion proposal, the unit's officers argued that they could provide more bang for the buck than their conventional counterparts in the overall Canadian Forces. "Equal if not greater political and military leverage can be achieved by employing a small SF force versus large mechanized battle groups," one officer wrote. "Vietnam taught us that conventional forces and methodologies cannot triumph in an unconventional, asymmetrical threat environment."

In 2003 recruiting material, the Canadian military claims that JTF2 is now a "world-class special operations force." That boast may be a little premature based on the results of the unit's activities in one war, but there is no question that JTF2 is moving towards the creation of a formalized special operations capability.

In outlining plans for JTF2's expansion and future, its officers note that while they still want to concentrate on the unit's original domestic counter-terrorism mission, they will also be placing emphasis on SOF skills, such as the ability to work with their foreign counterparts on the battlefield, as well as handling specialized roles, for instance, reconnaissance and surveillance, and combat search and rescue. For the latter missions, JTF2 personnel would be used to rescue Canadian POWs or pilots forced down behind enemy lines. The intention for the future is also to expand JTF2's parachute capability for SOF missions.

The commanding officer's priorities, according to JTF2 documents, also include continuing and maintaining the unit's relationship with three key SOF/counter-terrorism formations: the British SAS, Delta Force and the U.S. Navy's DevGru (SEAL Team Six).

In the midst of all these changes, JTF2 observers will be watching closely, both to see if the unit's previous successes and solid reputation can be maintained as it attempts to double its capabilities over the course of just several years, and to determine whether political oversight of the group is improved the next time it is called upon for an overseas mission.

DHD 81-053-

RIGHT: Counter-terrorism and hostage-rescue will continue to be JTF2's main focus in the future, according to its officers. (COURTESY JTF2)

BELOW: A JTF2 operator (in vest) is on hand to provide protection for Canadian Defence Minister John McCallum during a visit to Afghanistan in the summer of 2003.
(COURTESY COMBAT CAMERA)

ABOVE: *Joint Task Force Two has been on a recruiting drive as it tries to meet the Canadian government's order to double its capacity by 2006. (COURTESY JTF2)*

DHD 01

DHD02-289-01

ABOVE LEFT: *JTF2 operators rappel down the side of a building during hostage-rescue training. Over the years, JTF2 has had to modify its selection process to ensure it has enough recruits.*

ABOVE RIGHT: *A JTF2 soldier on winter exercises. The unit's training in winter and mountain warfare was one of the keys to its successes in Afghanistan.*

RIGHT: *JTF2 assaulters prepare to enter a room on a hostage-rescue training mission. (ALL PHOTOS COURTESY JTF2)*

DHC98-013-01

JTF2 operators fast-rope from a Griffon helicopter to "attack" terrorists during a training exercise. / ***BELOW:*** *Members of JTF2's dive team slip ashore during a maritime counter-terrorism exercise. The dive team conducts joint training with the SBS and SEALs.*

ABOVE LEFT: *JTF2 plans to develop niche areas in special operations in addition to its counter-terrorism role. (ALL PHOTOS THIS PAGE COURTESY JTF2)*

OPPOSITE PAGE: *SEALs practice boarding a ship during a training exercise. North American ports have been deemed potential targets for al-Qaeda. (COURTESY USN)*

Terrorism: The Everlasting War

10 Hamid Karzai appeared relaxed as he leaned out of his car window to shake hands with well-wishers who had crowded around his vehicle. The Afghan president was in a good mood, having just attended his brother's wedding in the family's hometown of Kandahar. The presidential motorcade of SUVs had stopped for a moment, and Afghans jostled each other for a glimpse of the country's pro-Western leader.

Karzai was shaking the hand of a small boy when an Afghan soldier appeared in the crowd, lifting an AK-47 to his shoulder and aiming at the president's vehicle and the people inside. The gunman, a former Taliban who had been recruited into the ranks of the government's military forces only two weeks earlier, managed to fire off several shots at the vehicle. Two rounds smashed into the president's car window, and flying glass fragments slightly wounded Governor Gul Agha Sherzai who was with Karzai.

At the first crack of gunfire, Karzai's bodyguards, members of the SEAL counter-terrorist unit, DevGru, sprang into action. "Get the hell out of here!" one of the SEALs screamed, as the presidential convoy tore off at high speed.

In the commotion, two Afghan men tackled the would-be assassin, wrestling him to the ground. At that point, a SEAL aimed his M4 and opened fire, killing Karzai's attacker as well as the two Afghans who had come to the aid of the president.

Back at the governor's guest-house, the SEALs stood close watch over Karzai, not knowing if the Taliban or al-Qaeda planned to follow up with an assault on the compound. Although a further assault never happened, and Karzai escaped the attack unscathed, the September 5, 2002, assassination attempt was just one more sign that the war on terror was entering a new phase.

Although the U.S.-led military onslaught on Afghanistan had been successful in disrupting al-Qaeda and overthrowing the Taliban regime, by the spring of 2003 both organizations were regrouping and infiltrating back into the country from their safe havens in Pakistan's Northwest Frontier. And while Western forces controlled Kabul, the situation outside the capital had deteriorated with warlords once again in command of large portions of the country. In short, the initial battle in the war on terror had been won, but the larger war was far from over.

Even inside Kabul, Islamic extremists had their sympathizers. On June 7, 2003, al-Qaeda struck at coalition troops in the city, killing four German soldiers and injuring 29 others. The men died when a suicide-bomber rammed a taxi loaded with several hundred kilograms of explosives into their bus. The troops had been on their way to the airport for a flight home to Germany.

In Kandahar, the Taliban had unleashed a campaign of terror against government supporters. Clerics who voiced approval of the Karzai regime were assassinated and police officers gunned down. In August 2003, a remote-controlled bomb blew up at a mosque, injuring another pro-government Mullah.

Special operations units had been conducting precision attacks in an effort to break the Taliban and al-Qaeda resurgence throughout 2003. On July 19, a U.S. SOF patrol near Spin Boldak waged an intense firefight with enemy forces, eventually calling in Apache attack helicopters for support. By the end of the battle, 24 Taliban had been killed.

Raids on weapons caches also continued to require SOF resources. For example, on June 9, U.S. special operations forces discovered three Blowpipe surface-to-air missiles near the town of Asadabad in eastern Afghanistan.

And although special operations forces had their successes, they also continued to suffer casualties. On June 26, SEAL Petty Officer First Class Thomas Retzer died from injuries he received during a firefight south of Kabul in the mountains along the border of Pakistan. Two other special operations soldiers who were with the 30 year old at the time were wounded.

Elsewhere, American SOF had been taking the fight to the enemy. The U.S. was re-establishing a military presence in the Horn of Africa to track down al-Qaeda's network in that region. Marines and Green Berets were in the former Soviet republic of Georgia to help train the military there in tactics for use against

Islamic guerrillas. In the Philippines, SEALs had helped train the country's armed forces for operations against Abu Sayuf, an al-Qaeda-linked terrorist organization.

Al-Qaeda, however, was not waiting for the enemy to come to it. Throughout 2003, the network continued to strike at overseas locations frequented by westerners. On August 5, a group affiliated with bin Laden's organization detonated a massive car bomb outside the Marriott Hotel in Jakarta, Indonesia, killing 16 people and wounding 150.

Several months before, al-Qaeda launched a series of attacks in Riyadh, Saudi Arabia, including a raid on the residential quarters of the Vinnell Corporation, a company supplying ex-U.S. Army officers to train the country's National Guard. Twenty-nine people were killed and 50 wounded when suicide attackers hit three compounds. The principal planner behind the attacks was believed to be Kahled Jehani, who had escaped from the U.S. assault on Tora Bora in December 2001.

These events have raised a key question: When will al-Qaeda strike next in North America? The answer, according to most intelligence officials, is that it's only a matter of time.

In June 2003, the U.S. government presented a report to the UN that predicted al-Qaeda would try to launch a chemical, biological or nuclear attack within two years, although it did not state whether such an attack would be on American soil. FBI officials have said they believe that bin Laden's terror network is capable of at least one more attack on the scale of September 11.

A May 2003 report on al-Qaeda's future potential produced by the Rand Corporation, an American defence think-tank, warned about an attack aimed at disrupting the U.S. economy. Bin Laden himself has talked about the need for "economic attrition" and al-Qaeda has credited its September 11 attacks for substantially weakening America's economic power. But more needs to be done, bin Laden told his followers in a December 2001 videotaped message. "The young men need to seek out the nodes of the American economy and strike at the enemy's nodes," he said.

Whether that means planting bombs on Wall Street or launching a devastating attack on other high-profile targets, such as a nuclear power plant or bustling harbor, is, of course, unknown. But it is likely that any such terrorist strikes could do serious damage to the North American economy.

In response to these concerns, some SOF and counter-terrorism units have begun making changes on several levels.

Some, for example, are refocusing their training to emphasize the use by terrorists of weapons of mass destruction. In Canada, JTF2 has been focusing on the maritime terrorist threat, conducting training against the possibility that freighters may be used to carry a WMD into one of the country's harbors. JTF2 conducted maritime counter-terrorism exercises in Halifax, Nova Scotia, in the summer of 2002, as well as in Victoria, British Columbia, in the spring of 2003.

It is also recognized that there is a need to take into account the new mindset of the enemy. Unlike terrorist acts of the past, in which taking hostages was seen as a prelude to negotiations for some kind of political goal, al-Qaeda operations are designed with only one outcome in mind: the death of the terrorists in an attempt to inflict maximum damage on the enemy. Al-Qaeda's actions are swift and devastating; there won't be time for SOF to rescue hostages from embassies or aircraft. As the September 11 attacks demonstrated, aircraft are not seized as negotiating tools, they are used as weapons.

With that in mind, some SOF have begun changing their tactics. U.S. Air Force tactical units whose job is to respond to an intrusion or takeover at America's nuclear missile silos have a new set of orders in the aftermath of September 11. Past tactics, which called for containing the situation and establishing communications with individuals who had gained access to a missile silo, are a thing of the past. Now, the first team on the ground is to immediately engage and eliminate intruders with maximum violence. According to Air Force Major General Timothy McMahon, such units will be brutal and overwhelming. There will be no hostage negotiations.

At the same time, the Air Force tactical units responsible for missile silo security will receive an estimated $1 billion dollars worth of new equipment, including a fleet of helicopters.

While various units train for what intelligence officials say will be an inevitable attack on North America, some of America's premier special operations forces are still tied down with the guerrilla war in Iraq. Delta Force and SEALs with Task Force 20 are focused on hunting for Iraq's elusive weapons of mass destruction, as well as trying to find Saddam Hussein. (The task force, backed by a quick-reaction force of conventional troops, killed his two sons in July 2003.)

Bush administration officials insist it is only a matter of time before Saddam is killed or captured, but there is concern that the Iraq war and ongoing operations in that country have diverted U.S. attention from battling terrorism, in particular Osama bin Laden's al-Qaeda network. Indeed, CIA operatives, as well as the U.S. Army's Fifth Special Forces Group, some 300 Green Berets in total, were reassigned from Afghanistan to the Iraq war.

Democratic Senator Bob Graham, a member of the Senate Intelligence Committee that investigated the September 11 attacks, believes the Bush administration lost its focus on the war on terror when it shifted its attention to overthrowing Saddam Hussein. That, according to Graham, allowed al-Qaeda to regroup and regenerate.

The attack on Iraq also caused dissent among some of the top counter-terrorism advisors at the White House. In the spring of 2003, Rand Beers, who served on the National Security Council as a special assistant to the president for combating terrorism, left his job over concerns that the Bush administration's policies had hurt the fight against al-Qaeda.

Beers, who had served as an advisor to Bush's father as well as to presidents Ronald Reagan and Bill Clinton, believes the U.S. abandoned the war in Afghanistan when it turned its attention to Iraq. He notes that the Bush administration not only failed to keep a significant military presence in Afghanistan, but it did not augment its military actions with diplomatic and aid efforts to deal with the root of terrorism.

This latter point is worth noting. After all, while special operations and covert forces can play a decisive role on the battlefield, past U.S. efforts have proven that covert operations cannot exist in a vacuum. If they are not accompanied by long-term foreign and strategic policies, they will ultimately cause more problems than they solve.

A prime example is the CIA's involvement in the overthrow of the democratically elected Iranian government in 1953. The spy agency helped install the Shah of Iran, whose abusive regime was overthrown in 1979 by Islamic extremists under the Ayatollah Khomeni. More than two decades later, the U.S. is still having to deal with the radical Islamic regime, which the Bush administration has named as part of the "Axis of Evil."

Then there was the decision – arguably a short-sighted one – in the 1980s to use American dollars along with CIA and SOF training to support the Mujahedeen war against the Soviets in Afghanistan. While that policy accomplished its goal of producing a Russian Vietnam, it also set the stage for the export of well-trained Mujahedeen to take part in guerrilla wars in places ranging from Algeria and Chechnya to Bosnia.

As Beers notes, a strong and continuing presence in Afghanistan is crucial. It is estimated that the destruction of al-Qaeda's infrastructure in the country prevented at least half a dozen of its terrorist missions from proceeding. It also disrupted bin Laden's research into weapons of mass destruction, which was in its early stages. But by turning its attention to Iraq, the U.S. has failed Afghanistan.

"Terrorists move around the country with ease," Beers said in an interview with the *Washington Post*. "We don't even know what's going on. Osama bin Laden could be almost anywhere in Afghanistan."

What's more, the invasion of Iraq appears to have become a rallying point for Islamic extremists around the globe. It is seen by such groups as proof of the secret U.S. agenda to dominate the Muslim world. By the end of the summer of 2003, U.S. generals suggested that many of the ongoing guerrilla attacks on American units, as well as bombings of key Iraqi installations, were the work of foreign Islamic terrorists intent on disrupting efforts to set up a Western-style democracy in the country.

Despite the criticism, the Bush administration appears confident it is winning the battle against terrorism. Half of al-Qaeda's senior leadership is either dead or behind bars. U.S. intelligence officials believe that most of bin Laden's fortune is gone, either used up to pay for operations or disrupted because of the financial crackdown on al-Qaeda accounts. Some 3,000 al-Qaeda operatives are in jails in various countries.

At a July 30, 2003, news conference, Bush spoke of the successes in the war and predicted the U.S. "will thwart" future al-Qaeda attacks.

Such declarations may one day come back to haunt the U.S.

Al-Qaeda has shown itself to be a patient organization that carefully selects its targets and painstakingly plans its missions. The 1998 bombings at the U.S. embassies in Kenya and Tanzania took five years to put together, while a similar length of time was used to plan and implement the September 11 attacks.

Furthermore, the Congressional investigation into the September 11 attacks concluded that anywhere between 70,000 and 120,000 terrorists have been trained by al-Qaeda over the years. Most have returned to their home countries, perhaps to carry out terrorist operations, while others have stayed in the U.S. as sleeper agents.

Navy SEAL Robert Harward, the Task Force K-BAR commander in Afghanistan and a key SOF operator in the Iraq war, has suggested that the new battle against terrorism is like nothing the world has ever seen. It is highly organized guerrilla warfare playing out on a global stage. In an interview with the *Ventura Country Star*, Harward warned that the public should be prepared for a very long, very non-traditional war.

"What people don't understand is we're dealing with a whole different enemy," said Harward. "Time is irrelevant to them. Targets are everywhere."

This elusive enemy may sit in the shadows for years, emerging only occasionally to strike a deadly blow when and where it is least expected. But the special operations forces of the West work in the shadows too – training, planning and adapting to the ever-changing rules of the new war against terrorism.

RIGHT: A Green Beret instructs new recruits in the Afghan army. The training program is an attempt to support the pro-western government of Hamid Karzai. *(COURTESY U.S. ARMY)*

BOTTOM: More SOF could be needed in the future in Afghanistan because of a resurgence of al-Qaeda and Taliban forces in that country. In this photo, an Afghan civilian barely notices an Australian SAS patrol driving by. *(COURTESY ADF)*

SELECTED BIBLIOGRAPHY

Connor, Ken. *Ghost Force: The Secret History of the SAS*, Cassell and Company, London, 1998.

Directory of the World's Weapons, Blitz Editions, Leicester, 1996.

Horner, David. *SAS Phantoms of War: A History of the Australian Special Air Service*, Allen and Unwin, Crows Nest NSW, 2002.

Katz, Samuel. *The Illustrated Guide to the World's Top Counter-Terrorist Forces*, Concord Publications, Hong Kong, 1995.

Moore, Robin. *The Hunt for Bin Laden*, Random House, New York, 2003.

Parker, John. *SBS, The Inside Story of the Special Boat Service*, Headline Book Publishing, London, 2003.

Prados, John. *Presidents' Secret Wars,* Elephant Paperbacks, Chicago, 1996.

Pugliese, David. *Canada's Secret Commandos: The Unauthorized Story of Joint Task Force Two*, Esprit de Corps Books, Ottawa, 2002.

Woodward, Bob. *Bush at War*, Simon and Schuster, New York, 2002.

Yousaf Mohammad, Adkin, Mark. *Afghanistan: The Bear Trap*, Leo Cooper, South Yorkshire, 1992.

NOTES ON SELECTED SOURCES

CHAPTER ONE

Qala-i-Jangi Uprising

House of War: The Uprising at Marzar-e-Sharif. CNN documentary, broadcast on August 3, 2002. As well, German journalist Armin Stauth has detailed his confrontation with SOF at the News Xchange journalism conference, October 10, 2002.

The Legend of Heavy D and the Boys, Robert Young Pelton, National Geographic Adventure, March 2002.

Five Americans hit by 'friendly fire' in battle to take prisoners in fortress, Associated Press, November, 27, 2001.

SAS are filmed firing in jail riot, Michael Evans, The Times, December 14, 2001.

British SBS commando to get American VC, Michael Smith, Daily Telegraph, March 26, 2002.

Inside the Battle at Qala-i-Jangi, Alex Perry, Time Magazine, December 1, 2001.

Soldiers tell of friendly fire bomb, Washington Post wire service, February 6, 2002.

Congressional Record: December 11, 2001, Honoring Johnny Micheal Spann, First American Killed in Combat in War Against Terrorism.

New Role for SOF

DoD May Boost Special Operations, Frank Tiboni, Defense News, July 2-8, 2001.

DoD (Department of Defense) briefing, transcript, October 18, 2001.

CHAPTER TWO

CIA operations and early days of the Afghan War

Yousaf Mohammad, Adkin, Mark. Afghanistan: The Bear Trap, Leo Cooper, South Yorkshire,

1992.

Woodward, Bob. Bush at War, Simon and Schuster, New York, 2002.

The CIA's Secret Army, Douglas Waller, Time magazine, February 3, 2003.

Report of the Joint Inquiry into the Terrorist Attacks of September 11, 2001, House/Senate Select Committee on Intelligence, December, 2002, partially declassified July, 2003.

Prados, John. Presidents' Secret Wars, Elephant Paperbacks, Chicago, 1996.

Green Berets/Combat Controllers

Moore, Robin. The Hunt for Bin Laden, Random House, New York, 2003.

Afghanistan and the Future of Warfare: Implications for Army and Defense Policy, Stephen Biddle, National Security Studies at the U.S. Army War College Strategic Studies Institute, Carlisle, PA, November, 2002.

An Unlikely Super-Warrior Emerges in Afghan War: U.S. Combat Controllers Guide Bombers to Precision Targets, Vernon Loeb, Washington Post, May 19, 2002.

Press Release: DoD Identifies Three U.S. Army Soldiers Killed, December 5, 2001.

Team 555 Shaped a New Way of War: Special Forces and Smart Bombs Turned Tide and Routed Taliban, Dana Priest, Washington Post, April 3, 2002.

Delta Force Raid on Mullah Omar's Compound

Escape and Evasion, Seymour Hersh, New Yorker, December 11, 2001.

Omar's Compound Was Raided, Airfield Also Hit In Rangers' Hunt For Intelligence, Vernon Loeb and Thomas E. Ricks Washington Post, October 21, 2001.

SBS/SAS/Delta Force/Tora Bora

Parker, John. SBS, The Inside Story of the Special Boat Service, Headline Book Publishing, London, 2003.

Commandos hit supply lines, Catherine Philp, The Times of London, November 24, 2001.

SAS forces the enemy back towards Kandahar, Michael Smith, Daily Telegraph, November 28, 2001.

Cave warfare: SAS take no prisoners in battle for caves, James Clark, The Times, December 2, 2001.

Wounded SAS men were in cave raid, Michael Smith, The Daily Telegraph, December 1, 2001.

Special Troops to Go After Al Qaeda, Steve Vogel and Vernon Loeb, Washington Post, December 14, 2001.

Is this the way bin Laden escaped?: How U.S. fear may have stopped the SAS from killing top terrorist leader, Bruce Anderson, National Post, February 16, 2002.

The Getaway, Seymour Hersh, The New Yorker, January 28, 2002.

DoD press briefing, transcript, Deputy Secretary of Defense Paul Wolfowitz, December 10, 2001.

CHAPTER THREE

Task Force K-Bar (Unclassified Briefing) Captain Bob Harward, July 8, 2002.

Mission was clear: Put 'eyes on target', James W. Crawley, San Diego Union Tribune, December 21, 2001.

Terror war led by special forces, James W. Crawley, San Diego Union Tribune, September 20, 2002.

Biography of Captain Robert Howard, provided by USN.

Hard-Shelled, SOF-Centered, Gordon T. Lee, Rand Review, Summer, 2002.

Commanders hail successes in Afghanistan, Dennis O'Brien, The Virginian-Pilot, October 3, 2002.

Zhawar Operation

On the Ground, How Special Ops Forces are Hunting al Qaeda, Mark Mazzetti, U.S. News and World Report, Feb. 25, 2002.

U.S. Aircraft Hit Zhawar Kili Complex Again, DoD press release, January 9, 2002 (Includes aerial photos and gun camera video).

Yousaf Mohammad, Adkin, Mark. Afghanistan: The Bear Trap, Leo Cooper, South Yorkshire, 1992 (details previous Russian assaults on the complex).

Hazar Qadam Raid

U.S. Was Misled in Deadly Raid, Afghans Say, John Fullerton, Reuters, January 31, 2002.

How 18 Afghan men were mistakenly killed by U.S. commandos, Preston Mendenhall, MSNBC Website, February 26, 2002.

Afghans killed at Hazar Qadam not Taliban or al-Qaida, Press release, Embassy of the United States, February 21, 2002.

DoD press conference, transcript, Donald Rumsfeld, General Richard Myers, February 21, 2002.

Operation Anaconda

Executive Summary of the Battle of Takur Ghar, DoD, May 24, 2002.

Results of Investigation into Death of U.S. Service Member Army Chief Warrant Officer Stanley L. Harriman, Unclassified Executive Summary, DoD, November 8, 2002.

Moore, Robin. The Hunt for Bin Laden, Random House, New York, 2003.

Affidavit of SSgt Kevin Vance, Bagram, Afghanistan, March 25, 2002.

DoD background briefing on the Report of the Battle of Takur Ghar, May 24, 2002

Operation Anaconda, interview with U.S. soldiers, DoD transcript, March 7, 2002.

General Defends Officers' Role in Afghan Battle, Robert Green, Reuters, May 24, 2002.

Bravery and Breakdowns in a Ridgetop Battle: 7 Americans Died in Rescue Effort That Revealed Mistakes and Determination, Bradley Graham, Washington Post, May 24, 2002.

'I died doing what made me happy': slain SEAL, Associated Press/Vancouver Sun, March 9, 2002.

Firefight on the Whale's Back, Stephen Thorne, Legion Magazine, May/June, 2003.

Afghanistan and the Future of Warfare: Implications for Army and Defense Policy, Stephen Biddle, National Security Studies at the U.S. Army War College Strategic Studies Institute, Carlisle, PA, November, 2002.

K-BAR Accomplishments

Statement to U.S. politicians by Commander Kerry Metz, Director of Operations, Task Force K-BAR, May 16, 2002.

Statement of Rear Admiral Joseph Krol, Assistant Deputy Chief of Naval Operations, House

Armed Services Committee Special Oversight Panel on Terrorism, June 28, 2002.

CHAPTER FOUR

U.S. Request for Australian SAS

Woodward, Bob. Bush at War, Simon and Schuster, New York, 2002.

Prior Relationship with SOCOM/SAS Vietnam Operations/Desert Thunder

Horner, David. SAS Phantoms of War: A History of the Australian Special Air Service, Allen and Unwin, Crows Nest NSW, 2002.

Operations in Afghanistan

Briefing, Brigadier Gary Bornholt, transcript, January 19, 2002.

Briefing, Brigadier Gary Bornholt, transcript, January 22, 2002.

Briefing, Brigadier Duncan Lewis, transcript, December 2, 2002.

Australian SAS in Afghanistan, Ian Bostock, Special Ops Journal of Elite Forces, Volume 21.

Inside the Australian anti-terror camp that isn't there, Craig Nelson, Sydney Morning Herald, February 16, 2002.

SAS men who dared and won decorations, Ian McPhedran, Courier-Mail, November 28, 2002.

Radio interview SAS Signaler Martin Wallace, Australian Broadcasting Corporation transcript, November 27, 2002.

Maj. Gen. Duncan Lewis: Special Operations Commander Australia, Ian Bostock, Jane's Defence Weekly, May 28, 2003.

Operation Condor

Marines tell of major battle. MoD say they have not engaged the enemy. Who is right? Alex Spillius, Michael Smith, The Telegraph, May 18, 2002.

Al Qaeda hunt: violent shell game, Scott Baldauf, The Christian Science Monitor, May 20, 2002.

CHAPTER FIVE

(all Department of National Defence (DND) JTF2 documents have been declassified)

Task Force K-Bar (Unclassified Briefing) Captain Bob Harward, July 8, 2002.

More JTF2 commandos in war zone, David Pugliese, Victoria Times Colonist, December 20, 2001.

Pugliese, David. Canada's Secret Commandos: The Unauthorized Story of Joint Task Force Two, Esprit de Corps Books, Ottawa, 2002.

JTF2 Training Missions

Various JTF2 Historical Reports

JTF2 POW Affair

Op Apollo Interim Prisoners of War Handing, Briefing Note for the Chief of the Land Staff, January, 17, 2002.

Issues Related to Persons Detained by the Canadian Forces and to Canadian Forces Members Detained by Opposing Forces, Department of National Defence, January, 2002.

Liberals in Afghan Muddle, Confusion leads to uproar in House, Sheldon Alberts, National

Post, January 30, 2002.

Interview Deputy Prime Minister John Manley, transcript, February 5, 2002.

POW chronology, prepared by DND at request of PCO, February 20, 2002.

Transcripts, Standing Committee on Procedure and House Affairs, February 20/26/27, 2002.

JTF2 Dealings with News Media

Various interviews of Canadian journalists assigned to Kandahar provided insight into JTF2 at the base.

Commander's Assessment Report, DND, February 6, 2002 (details various problems with news media).

Canadian troops expel reporter from airbase, Daniel Leblanc, Globe and Mail, February 13, 2002.

His secret is out: New commander was in commando unit in terrorism war, Paul Cowan, Edmonton Sun, August 16, 2002 and Stogran turns over battalion command, Keith Gerein, Edmonton Journal, August 16, 2002. (The problem of JTF2's supposedly iron-clad secrecy policy is amply displayed by the these two articles. Both reveal that Mike Beaudette served with JTF2 in Afghanistan. The media found out about his service with the unit after it was alluded to in a public speech by a retired senior military official at a Change of Command parade involving Beaudette).

JTF2 Equipment Issues

Naval Special Warfare Expands Command Role in Joint Force, Roxana Tiron, National Defense, February, 2003.

Deployment of Special Operations Forces Variant HMMWV to Afghanistan, Briefing Note for Chief of the Defence Staff, August 30, 2002.

Sole Source Procurement JTF2, Briefing Note for Deputy Chief of the Defence Staff, January 31, 2002.

Mirwais Hospital Siege

Six al-Qaeda killed at hospital siege, BBC News, January 28, 2002.

Shootout involved Canadians, Mark MacKinnon, Globe and Mail, February 7, 2002.

A Bulletproof Mind, Peter Maas, New York Times Magazine, November 10, 2002.

Dead Men Talking, Michael Ware, Time, February 2, 2002.

Moore, Robin. The Hunt for Bin Laden, Random House, New York, 2003.

Operation Anaconda

CENTCOM Press Briefing, Operation Anaconda, General Tommy Franks, March 4, 2002.

JTF2 locked in fierce Afghan fight: Government releases few details about battle. Tim Naumetz, David Pugliese and Hilary Mackenzie, Ottawa Citizen, March 5, 2002.

Band Taimore Village Raid

The Allies are still on the hunt in Afghanistan—and the locals aren't happy, Michael Ware, Time Magazine, June 24, 2002.

Afghan Villagers Angry Over US Raid, Patrick Quinn, Associated Press, May 26, 2002.

DoD press conference, Donald Rumsfeld, May 30, 2002.

Tracking down the Taliban, but at what cost?, Mike Vernon, CBC News website, June 3, 2002.

Coalition Raid Conducted in Afghanistan, DND press release, May 24, 2002.

Military releases Afghans snared by commandos, Nahlah Ayed, Canadian Press, May 31, 2002.

Interviews by author - SOCOM/DND officials.

JTF2 Returns from Afghanistan

Shadowy JTF2 remains behind to stalk enemy, Canadian Press, July 25, 2002.

From the Desk of Defence Minister John McCallum, Maple Leaf (Newspaper of the Canadian Forces/DND), August 28, 2002.

JTF2 Recruiting and Information Video, Department of National Defence, 2003 version.

Honours and Awards Process for Special Operations Forces, Briefing Note for the Chief of the Defence Staff, December 10, 2002.

CHAPTER SIX

A new terror-war front: the Caucasus, Fred Weir, The Christian Science Monitor, February 26, 2002.

Peru Hostage Crisis; Personal Update, declassified (Anthony Vincent), April 18, 1997.

Fifth Column Among Hostages, Sergey Stefanov, Translated by Maria Gousseva, Pravda, Oct, 29 2002.

Russia's Special Operations Forces, (Colonel, retired) Stanislav Lunev, Translated by Mark Eckert, Prism.

Transcript, State Department Comments on Hostage Crisis at Moscow Theater, 28 October 2002.

Gas Killed most Hostages, Carolynne Wheeler, Ottawa Citizen wire services, October 28, 2002.

Moscow Rescue Descended Into Bedlam, Olga Nedbayeva, Agence France Press, October 31, 2002

What Have They Done to Us, Matthew Fisher, National Post, November 2, 2002.

This Was Our Sept. 11, Christina Lamb, Sunday Telegraph, October 27, 2002.

Russia's Mystery Gas, Preliminary Briefing Note, Assistant Deputy Minister for Policy, DND, October 30, 2002 (declassified).

Terrorist Leader Died Drinking Cognac, Pravda, translated by Maria Gousseva, October 26 2002.

A Russian spy's inside job, Mark Franchetti, The Times of London, November 3, 2002.

Picture Emerges of How They Did It, Nabi Abdullaev, Moscow Times, November 6, 2002

CHAPTER SEVEN

Australian SAS

Australian SAS Trooper Awarded Medal for Gallantry, Australian Defence Force news release, May, 18, 2003.

Edited version of the Trooper X citation, May 18, 2003.

Colonel John Mansell, Australian Special Forces Contribution to Operation Falconer (Iraq), speech to RUSI, May 9, 2003.

Australia put troops into Iraq before war, Associated Press, May 10, 2003.

SAS troops in front line against suicide bombers, Tom Allard, The Age Company Ltd, April 2 2003.

Diggers help secure town, Mark Forbes, The Age Company Ltd, April 17, 2003.

Half of Saddam's combat fighter force unearthed Neil Tweedie, Toby Harnden, The Telegraph, April 19, 2003.

Franks' War Plan

Allied Special Forces Took Western Iraq, Vago Muradian, Defense News, May 19, 2003.

Iraq's western desert a 'special forces playground' Tim Ripley, Jane's Defence Weekly, April 9, 2003.

The New War Machine, Peter Boyer, The New Yorker, June 30, 2003.

U.S. Concerns about al-Qaeda link to Saddam Hussein

Woodward, Bob. Bush at War, Simon and Schuster, New York, 2002.

Al Faw

Secret Armies of the Night, Michael Duffy, Mark Thompson and Michael Weisskopf, Time magazine, June 23, 2003.

Unheralded copter pilots on carrier play key role, Otto Kreisher, Copley News Service, March 24, 2003.

SEALs give glimpse of missions in Iraq James W. Crawley, San Diego Union Tribune, June 27, 2003.

Northern Iraq

Peshmerga interviews conducted by Scott Taylor, Northern Iraq, May, 2003.

U.S. Troops Working With Kurdish Fighters Groups May Help Special Forces Plan Airstrikes for Advance Into Northern Iraq, Daniel Williams, Washington Post, March 17, 2003.

Northern Iraq Firefight, Victor Black, Soldier of Fortune, July, 2003.

'The US bomb fell just 10 yards from me, minutes later my translator was dead', John Simpson, The Telegraph, April 7, 2003.

Two SBS survive ambush, desert trek and Syria jail, Michael Evans, The Sunday Times, April 29, 2003.

Helicopter pulls out SAS team after secret mission uncovered, Neil Tweedie, The Telegraph, April 3, 2003.

Murkaryin Dam assault

Special units played bigger role this time, James W. Crawley, San Diego Union-Tribune, June 27, 2003.

Nick Withycombe (Australian SAS)

Back from the front line, a soldier shares his tale, Paul Daley, Sydney Morning Herald, May 12, 2003.

British SAS/CIA/Delta Force/ Task Force 20

SAS joins US forces hunting for dictator, Michael Smith, Oliver Poole, The Telegraph, May 11, 2003.

Special forces inside capital as bombs black out power supply, Michael Smith, Neil Tweedie, The Telegraph, April 4, 2003.

Body in air raid debris is Chemical Ali, say Iraqi police, Neil Tweedie in Qatar and Robin Gedye, The Telegraph, April 8, 2003.

Who the **** are you, asked the man from special forces, Olga Craig, The Telegraph, April 6, 2003.

Missile base seized by SAS, Robert Winnett, Paul Ham, Times Newspapers Ltd., March 30, 2003.

SAS catch Saddam's half-brother, Michael Smith, The Telegraph, April 19, 2003.

Special Search Operations Yield No Banned Weapons, Barton Gellman, Washington Post, March 30, 2003.

Covert Unit Hunted for Iraqi Arms, Barton Gellman, Washington Post, June 13, 2003.

Grom

The GROM Factor - Haven't heard of Poland's Special Forces? They're real, they're serious, and they're here to save the day, Victorino Matus, The Weekly Standard, May 8, 2003.

Polish forces criticized for flag photos, Jan Repa, BBC News Online, March 25, 2003.

GROM makes us proud, Statement Polish Embassy, Washington, May 21, 2003.

Private Lynch

Spec Ops say lives were on line in Lynch's rescue, Rowan Scarborough, Washington Times, June 9, 2003.

The reality behind the myth, Hugh Dellios, E.A. Torriero, The Times, May 29, 2003.

SOF Successes/Losses

Special Ops steal show as successes mount in Iraq, Rowan Scarborough, The Washington Times, April 7, 2003.

Special Forces come of age, James Dao, New York Times, April 29, 2003.

Two wounded U.S. special ops forces rescued in Iraq, Reuters, April 9, 2003.

Operation Iraqi Freedom - By the Numbers, USAF, May, 2003, unclassified report.

CHAPTER EIGHT

Quiet Professionals, General Charles Holland USAF, Armed Forces Journal International, February, 2002.

DoD press conference, Donald Rumsfeld, transcript, January 7, 2003.

Backgrounder briefing with Senior Defense Officials on Special Operations, DoD transcript, January 7, 2003.

Special Ops Forces Are 'Tool of Choice,' David Litt, National Defense Magazine, February 2003.

Navy plans to expand its special forces, Otto Kreisher, Copley News Service, June 6, 2003.

Wanted (badly): more Green Beret recruits, Patrik Jonsson, The Christian Science Monitor, February 21, 2002.

Army recruits now enlisting into Special Forces, Spc. Kyle J. Cosner, Army News Service, April 12, 2002.

U.S. Marines to join Spec-ops World, C. Mark Brinkley, Defense News, June 17-23, 2003.

Army enacts 'stop-loss' for some specialties, Army News Service, December 4, 2001.

Warriors training warriors, Glen Goodman Jr., Armed Forces Journal International, September, 2002.

Special Operations Special Report, Gregor Ferguson, David Pugliese, Barbara Opall-Rome, Frank Tiboni, Jason Sherman, Defense News, April 21, 2003.

CHAPTER NINE

(all Department of National Defence (DND) JTF2 documents have been declassified)

JTF2 Expansion/Training Standards

Pugliese, David. Canada's Secret Commandos: The Unauthorized Story of Joint Task Force Two, Esprit de Corps Books, Ottawa, 2002.

JTF2 Annual Historical Reports, 1994-2001, DND.

Anti-terrorist unit to expand, Jim Bronskill and David Pugliese, Vancouver Sun, December 11, 2001.

Briefing Note, JTF2 Capability Enhancement, 1996.

JTF2 Recruiting and Information Video, Department of National Defence, 2003 version.

JTF2 Level 2 Business Plan 2001/02.

Board of Inquiry Change of Command, JTF2, June 14, 2000.

JTF2 Status and Future Growth Briefing to Armed Forces Council, January 23, 2001.

JTF2 Expansion Proposal, October 1, 2001 (draft).

Creation of CTSO Directorate, December 20, 2001.

Meeting Record and Decision Sheet - JCRB 02/02.

Project Charter, Canadian Forces Counter Terrorism and Special Operations Enhancement, March 20, 2002.

Canadian Forces School of Search and Rescue, recruiting supplement, 2003.

CANFORGEN JTF2 Selection, 1994 -2003.

Ways of Improving the selection procedure for special forces personnel, LTC Gunter Kreim, Maj. Rene Klein (unclassified report looks at German KSK acceptance rates).

Roles/Future Expansion JTF2

The Role of Special Forces in the Achievement of Foreign Policy Objectives, Maj. B.J. Brister, March 8, 2002.

The Future of Counter-Terrorism and Special Operations Forces in Canada, Major David Last, May 1999.

Interviews by author, Lieutenant-General Mike Jeffery, Brigadier General Vince Kennedy, Colonel Bill Peters, Colonel Howie Marsh, conducted during teleconference, May 9, 2002.

Overseas Operations

Op Stable, JTF2 Report

Op Dubonnet, JTF2 Report.

COP Zombie, JTF2 Report

JTF2 After Action Reports, 1996-2000.

Political Oversight

Briefing Note for Minister of National Defence, Tasking of JTF2, April 20, 1999.

Elite U.S. unit keeps heat on terrorists, July 12, 2002 (details briefings to President Bush).

Connor, Ken. Ghost Force: The Secret History of the SAS, Cassell and Company, London, 1998 (details SAS accountability).

CHAPTER TEN

Karzai Attack

Karzai survives shooting, Peter Goodspeed, National Post, September 6, 2002.

SEAL Team Six in Kabul: Eliminate the Threat, Robert K. Brown, Soldier of Fortune, December 2002.

Al-Qaeda Attacks/Afghanistan Situation

UN warned over Afghan 'time bomb', BBC News Service, June 18, 2003.

Bomb suspects fought in caves at Tora Bora, Robin Gedye, Adel Darwish, Ottawa Citizen, May 15, 2003.

Kabul Loses Sense of Safety After Killing of Peacekeepers, April Witt, Washington Post, June 15, 2003.

Future Attacks/New Tactics

Al-Qaeda, Trends in Terrorism and Future Potentialities: An Assessment, Bruce Hoffman, Rand Corporation, May 2003.

Protecting U.S. nuclear missile sites requires tougher, clear cut responses, William Scott, Aviation Week and Space Technology, May 12, 2003.

The Islamic Jihadist Threat after Afghanistan, April 2002, Directorate of Strategic Analysis, Canadian Department of National Defence (declassified).

U.S. Sees Likely Al Qaeda WMD Attack Within 2 Years, Reuters, June 9, 2003.

JTF2 readies for naval battle: Unit beefs up skills to fight terrorists in ports, high seas, David Pugliese, Ottawa Citizen, September 22, 2002.

Concerns About Iraq

Former Aide Takes Aim at War on Terror, Laura Blumenfeld, Washington Post, June 16, 2003.

US shifting focus, agents from Kabul to Baghdad, Bryan Bender, Boston Globe, August 18, 2003.

Future

Prados, John. Presidents' Secret Wars, Elephant Paperbacks, Chicago, 1996.

Statement by General Charles Holland to Senate Armed Services Committee, Emerging Threats and Capabilities Subcommittee, March 12, 2002.

Navy SEALs Led Covert War Against Al-Qaida, Seth Hettena, Ventura County Star, September 22, 2002.

198 SHADOW WARS: Special Forces in the New Battle Against T

INDEX

Abu Sayuf group: 183. *See also* al-Qaeda
   attacks in the Phillipines: 117
Afghan soldiers: **65**
Afghan tribes
   Mangal: 93
   Sabri: 93
Afghani allies
   manipulation of SOF: 45
Afghanistan
   Americans enter: 19
Air strikes: 23
Aircraft
   A-10s: 138
   AC-130 gunships: 90, 136
   Antonov cargo planes: 84
   Apache (attack helicopters): 38, 52
   B-1B bombers: 138
   B-52 bombers: 90, 142
   C-130 transports: **18**, 87
   CH-53 helicopters: 106
   Chinook helicopter: **97**
   EC-130 electronic warfare: 136
   F-14 Tomcats: 144
   F-18 fighter jets: 142
   Griffon helicopter: **180**
   Harrier jets: 138
   MC-130E Combat Talon: 146, 152, **165**
   crash of: 61
   MH-47 Chinook helicopter: 88, 138
   MH-53 Pave Low helicopter: 23, 38,
   77, 103, 106, 135
   P-3 Orion maritime patrol: 40, 135
   Pave Hawk helicopter: **159**
   Predator UAV: 37, 40, 54, 135
   SH-60 Seahawk: 136
   Spectre gunships: 142
   U.S. Black Hawk
     medevac: 86
al Asad airbase (Iraq): 148
   equipment, weapons found: 148, 148–
   149
Al Faw, Iraq: 135
al Zarqawi, Abu Musaab: 141

al-Islam, Ansar: 141
al-Jazeera (news agency): 25, 124, 144
al-Majid, "Chemical" Ali Hassan: 147
al-Qaeda: 8, 9, 14, 19, 20, 21, 23,
   25, 28, 29, 30, 31, 32, 33, 34,
   38, 40, 41, 42, 43, 44, 45, 46,
   47, 48, 51, 52, 53, 54, 55, 56,
   57, 58, 59, 60, 61, 82, 83, 84,
   85, 86, 87, 88, 89, 90, 91, 92,
   93, 94, 95, 99, 103, 104, 105,
   106, 107, 108, 110, 111, 112,
   117, 118, 121, 122, 130, 133,
   141, 142, 143, 144, 160, 182,
   183, 184, 185, 186
   attack at Bali, Indonesia: 117
   attack on American embassies (Africa): 8
   attack on World Trade Center (1993): 8
   attacks coalition troops (June 2003): 182
   begins to return to Afghanistan: 182
   complex at Zhawar: **63**
   connections to Ansar al-Islam: 141
   fighting at Tora Bora: 31–32
   in Chechnya: 121
   in Horn of Africa: 182
   in Pankisi Gorge, Georgia (Soviet republic
     of): 130
   raid on Vinnell Corporation: 183
   situation in 2003: 186
al-Sahhaf, Mohammed Saeed: 145
al-Tikriti, Watban Ibrahim Hasan
   captured: 148
Ali, Hazrat: 32
Ali Kheyle, Afghanistan: 47–48
Ammash, Huda Salih Mahdi: 150
Ansar al-Islam (terrorist organization). *See
   also* Terrorist forces
ar-Rutba, Iraq: 138
Armed Forces Journal Internationa (newspaper): 162
Associated Press, The: 103
Australia
   4th Battalion, Royal Australian Regiment
     (Commando): **74**, **158**
   in Operation Iraqi Freedom: 132
   commitment to War on Terror: 82–83, 84
   Special Air Service (SAS): 38, 48, 56,